GOVERNING
BEHAVIOR

GOVERNING
⌈BEHAVIOR⌉

*How Nerve Cell Dictatorships and
Democracies Control Everything We Do*

ARI BERKOWITZ

Harvard University Press
Cambridge, Massachusetts
London, England
2016

First printing

Library of Congress Cataloging-in-Publication Data

Berkowitz, Ari, 1961– author.
    Governing behavior : how nerve cell dictatorships and democracies
control everything we do / Ari Berkowitz.
        pages   cm
    Includes bibliographical references and index.
    ISBN 978-0-674-73690-0 (cloth : alk. paper)
        1. Neurons.   2. Neural networks (Neurobiology)   3. Animal behavior.
    I. Title.
    QP363.3.B47   2016
    612.8'1046—dc23   2015033365

In memory of
PAUL GROBSTEIN
&
JON I. ("JACK") ROBERTS

Contents

Color illustrations follow page 118

GOVERNING
BEHAVIOR

How to Spy on the Government

One morning, you get behind the wheel of your car and become lost in thought. The next thing you know, you arrive at your child's school, when you intended to drive to the grocery store.

What happened?

Something made you do a very complex series of things—including pressing the accelerator and the brake at the appropriate times, steering left and right as needed, signaling, turning, avoiding other vehicles, and so on—in a very precise way that brought you to a familiar location—all without your being aware of it.

What is the "something" that made you do it?

Your nervous system is the culprit.

Each movement you make, whether simple or complicated, is caused by muscle contractions, specifically contractions of your skeletal muscles, which are the only muscles you have voluntary control over.

Your skeletal muscles don't contract by themselves. They take orders. The orders come from a particular group of nerve cells, or neurons, called motor neurons. Whether your knee jerks in response to a doctor's tap or you perform a Rachmaninoff piano concerto on the Carnegie Hall stage, your skeletal muscles carry out the orders of motor neurons.

So do motor neurons rule our behavior?

Not really, because motor neurons themselves receive instructions from a variety of other neurons that make up your nervous system. Some of these instructions are excitatory—they say, "make the muscle contract." Others are inhibitory—they say, "stop the muscle from contracting." Each motor neuron adds up all these instructions—which can be thought of as votes—and determines the election result: contract or don't contract this muscle at this moment. This kind of voting is going on all the time in all the motor neurons that control all your skeletal muscles.

But this kind of voting doesn't just occur in motor neurons. It occurs constantly in all the neurons of your nervous system, some of which send their own results to motor neurons, and it determines whether each of your muscles contracts or not at each moment. In this way, your nervous system governs all your behaviors.

This doesn't mean that you don't have free will. You may say, I thought long and hard about my career path, or about which cereal to eat this morning, and I came to a logical conclusion based on the best available evidence. And that may be true. But it is also true that all of this deliberation, as well as the eventual movement of your hand toward the cereal box, occurred by way of signals sent among your neurons.

At every moment, each of your billions of neurons has to "decide" whether to send a signal to its own particular "recipient list" of other neurons. This decision is made by adding up excitatory and inhibitory inputs from many other neurons, sometimes thousands of them. This happens constantly throughout your life, at least when you're awake. This is equally true whether you are aware of your behavior or not.

Suppose you are walking with a friend along a path, engrossed in a discussion about Congress's inability to agree on anything or a new YouTube video. Your legs move back and forth, propelling you along at exactly the right speed to keep up with your friend. Your body maintains a posture that keeps you from falling. You angle your strides to follow the curved path. You skillfully avoid a low tree branch and step over a tree root. And you can do all of these things even without thinking about them, thanks to your nervous system.

If everything we and other animals do is regulated by the nervous system, we might think of the nervous system as a kind of government. If so, what form of government is it? Are there "dictator neurons" that command each movement? Or are there "democracies of neurons" that vote on our next move? What exactly is going on in there?

These are difficult but not impossible questions to answer. To answer them, you might say we have to spy on the nervous system. For this kind of spying, we have to monitor the electrical signals that neurons produce to trigger behaviors.

Neurons generate their own special kind of electrical signal, called an action potential or "spike." Spikes are the way that neurons represent and transmit information. Spikes are sudden but brief changes in the voltage inside a neuron (see Figure 1.1). The voltage jumps up and then back down in roughly 1 millisecond (one one-thousandth of a second). A spike is triggered whenever the voltage inside the neuron increases (we say the neuron gets "excited") beyond a certain threshold. This usually happens in or near the main part of a neuron, called its cell body, where most of the ordinary cellular functions occur.

Spikes have an unusual feature: after being triggered at the cell body, they can "travel" along another part of the neuron, called its axon, in something like the way electrons move down a wire. An axon is typically long and narrow—as

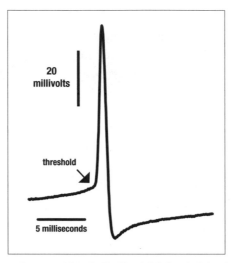

Figure 1.1. An action potential, or "spike," is a brief electrical signal that neurons generate to communicate with other cells. (This spike was generated by a turtle spinal cord neuron.)

long as the distance from your spinal cord to your foot—analogous to a wire or cable. (See Figure 1.2 for examples.) What actually happens when a spike travels along an axon, though, is that each spike triggers another in the next segment of axon, similar to one domino knocking over the next within a long series of upright dominoes.

Spikes in one neuron can also trigger spikes in other neurons through connections between neurons called synaptic connections or synapses. Through synapses, signals are usually transmitted from the end of the axon of the first neuron to branch-like structures called dendrites on the second neuron. (See Figure 1.2 for examples.) The dendrites, like the axon, grow from the cell body.

A whole set of synaptically connected neurons is in some ways like an electrical circuit that you might find inside elec-

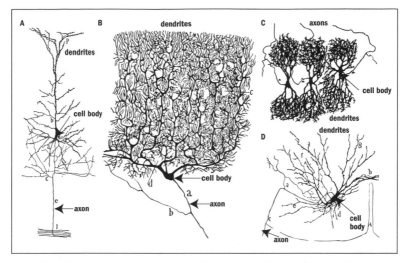

Figure 1.2. Examples of neurons, from the drawings of Santiago Ramon y Cajal, a founder of neurobiology. In each case (A–D), the cell body, dendrites, and axon are indicated. The cell body contains the cellular machinery that is found in all cells; the dendrites receive most electrical inputs from other neurons; the axon is how a neuron transmits electrical signals called spikes over long distances.

tronic equipment, so biologists often refer to a set of connected neurons as a "neuronal circuit." Within a neuronal circuit, each neuron typically makes synaptic connections to many other specific neurons—its recipient list.

I will argue that the best way to find out how nervous systems govern behaviors is to monitor spikes in the relevant neuronal circuits while behaviors are being produced. Spikes in these circuits directly cause motor neurons to spike, which in turn causes muscles to contract, generating particular forces at particular times. The whole set of muscle contractions makes up a behavior. So spikes in these circuits cause behaviors.

When we talk about the cause of an event, we can mean many different things. We can recognize, for example, that neurons are only able to spike and muscles are only able to contract because of the particular kinds of molecules within them and the ways that these molecules are arranged. In other words, the material and the form of each cell cause behavior.[1] This is certainly true.

Behaviors can also be explained by the history of changes in DNA that occurred over countless generations as each organism was altered from a previous version. The forms of organisms that were most successful and prolific in their environments are the ones still here. The sets of DNA that generate successful organisms lead to particular kinds of animal behaviors. Behaviors that contribute to survival and reproduction are thus passed down through the generations. In this sense, a particular evolutionary history is the cause of a behavior.[2] The evolutionary cause of a behavior is often called its ultimate cause.

Still another kind of cause of each event is whatever occurs immediately before the event and triggers it to begin. This is often called the proximate cause.[3] In the case of behaviors, a set of electrical events—spikes—in neurons is the proximate cause of any behavior. I argue that the proximate cause has a special significance for our understanding of how things happen, so we must find out about spiking in the neurons connected to the relevant motor neurons to fully understand the cause of any behavior.

To see the importance of knowing the proximate cause, let's consider another analogy. Suppose someone told you that a friend of yours was in a car accident. What would you say? First, you would probably want to know if your friend is okay. Next, you would probably ask, "What happened?!" What kind of an answer would you expect?

Probably, you would expect to hear the sequence of events that led to the collision of your friend's car with something else. Perhaps a drunk driver hit your friend's car, or perhaps your friend swerved to avoid an animal crossing the road and then collided with a guardrail.

If you were told instead that the car accident occurred because the car was made of heavy steel (the material), or because cars are constructed so that they cannot stop suddenly (the form), you would probably not be satisfied. If you were told that the accident occurred because cars and roads have been modified over generations to allow cars to travel at increasingly dangerous speeds (the evolution), you would probably be impatient for a different kind of answer. The composition and history of cars and roads are doubtless important contributors to dangerous car accidents in general, but they are things that we take for granted. They are indirect causes that don't explain what happened during this particular accident.

You would also want to know the direct or proximate cause of the collision—the series of events that led up to the collision and so "connects the dots." When it comes to behaviors, the proximate cause always involves electrical signals in neurons that activate a set of muscles, causing the muscles to contract in a particular sequence and to generate particular forces. So if we want to know what caused a certain behavior in a human or other animal, we have to find out which neurons spiked and when, to generate those muscle contractions.

In the process of studying which neurons spiked and when, we can investigate how the nervous system is organized to trigger this particular behavior in this particular situation. Essentially, by studying the electrical signals in neurons, we can study the way nervous systems govern behaviors.

Not all nonhuman animals are equivalent when it comes to studying how neuronal circuits control behaviors, for practical reasons. In general, neuronal circuits are much easier to investigate in smaller nervous systems, because there are fewer neurons and synaptic connections to map. In many invertebrate nervous systems there are individually identifiable neurons that can be named and located from one animal to the next within the species. Each identified neuron generally makes the same synaptic connections with other cells and is active in the same way during the same behaviors. This is a tremendous advantage for working out neuronal circuitry.

The optimal species for studying how a particular behavior is controlled often are *not* the very few species—such as rats and mice—that get much of the attention in biology today.[4] Instead, they may be fascinating species that have evolved special capabilities.

Every animal has a nervous system that controls its behavior. Different species are specialists or "champions" at doing different things. If you want to study, for example, how hearing or smelling is accomplished, it helps to study an animal that has a superior sense of hearing or smell. Their neuronal circuits are likely easier to identify and map, because their nervous systems emphasize auditory or olfactory behaviors and have developed these circuits to operate faster, more precisely, or with a clearer organization.

Still other species have particular technical advantages. For example, they might have very large neurons, which makes it easier to monitor their electrical signals. Or they might have a nervous system that stays healthy and functions normally for a long time, even when it is placed outside the animal, in a dish of saline.

This general idea was argued in 1929 by a physiologist named August Krogh, who wrote, "For such a large number of problems there will be some animal of choice or a few such animals on which it can be most conveniently studied."[5] This is now known as Krogh's Principle.

For example, in the 1930s and 1940s, Alan Hodgkin and Andrew Huxley discovered how neurons produce spikes. But Hodgkin and Huxley did not use mice to figure this out. Instead, they performed their experiments on *squid* giant axons (*not* giant squids, but giant axons of ordinary squids). They used these animals because a squid giant axon is so wide that you can insert a narrow glass tube into it by hand without destroying it.

Over the past few decades, neurobiologists (in this case also called neuroethologists) have been able to apply Krogh's Principle to address all kinds of fascinating questions about how neuronal circuits govern behaviors. For example:

Are behavioral decisions made by a single "dictator" neuron or by a "democracy" of many neurons, whose votes are tallied to determine the outcome?

How do animals run the "factories" of our bodies that produce rhythmic behaviors that are key to life, such as breathing, walking, swimming, and flying? Do nervous systems rely on central nervous system ("government") programs to plan in advance how to meet the body's needs, or do they instead rely on sensory feedback ("free-market") mechanisms?

How do nervous systems acquire intelligence on the outside world? For example, how do owls and echolocating bats use sound, how do star-nosed moles use touch, and how do weakly electric fish use electric fields to create high-resolution "reconnaissance images" of their surroundings?

How do nervous systems keep track of their own decisions?

How do animals attract partners or allies and deter rivals—what we might call playing politics? What happens in the nervous system when birds learn to become political animals?

What can we learn about how our own nervous systems govern behavior by studying how other animals do it?

We will turn to these topics soon.

2

Isn't There an Easier Way?

Monitoring spikes in neuronal circuits in a living animal is very hard to do. It is especially hard to monitor electrical signals from inside a neuron. To do this, you usually have to poke a sharp glass pipette filled with saline (called a microelectrode) into a neuron without destroying it. (You can also monitor spikes from just outside a neuron, which is somewhat less challenging but usually provides no information about electrical events in the neuron that are below the spike threshold.) Isn't there an easier way to find out the causes of behaviors? And naturally, we all want to know what causes *human* behaviors, not just those of other animals. Monitoring spikes in neurons throughout human brain circuits is actually a major goal of President Barack Obama's recent Brain Research through Advancing Innovative Neurotechnologies (BRAIN) Initiative, but we don't yet have technology that can do this noninvasively.[1] In most cases, we would not poke electrodes into a person's brain while they are doing something, for ethical reasons. (But stay tuned for exceptions to this rule.) Isn't there a suitable way to find out the proximate causes of human behaviors without poking electrodes into human brains?

In the popular media, human behaviors are rarely explained in terms of the electrical activity of neurons that trigger and generate behaviors. We simply don't have that kind of information for most human behaviors, because of the technical and ethical limitations. So how are human behaviors and behavioral differences usually explained? One common approach is to link a particular human behavior or behavioral variant with a certain DNA sequence within the person's genome. Another common approach is to link a behavior to increased physiological activity in one or more human brain areas, as assessed by conducting a brain scan while a person performs a particular task. Both of these kinds of research are valuable and connect aspects of human biology with human behaviors. But neither provides a direct or proximate explanation of the behaviors. Before we look at studies of spiking in neuronal circuits during behaviors, let's take a brief look at each of these better-known approaches to see why it's still necessary to investigate spiking in neurons if we want to know how nervous systems govern behavior.

Genes and Behaviors

Genetic explanations of behavior, or more commonly explanations of differences between the behaviors or abilities of different people, hinge on finding correlations between a form of the behavior and a version of a particular gene or DNA sequence.

Each gene is a sequence of components within a long DNA molecule, specifying production of an RNA molecule ("transcription") via a complementary sequence. Some RNA molecules have their own direct function, for example, to alter protein synthesis from other genes. Other RNA molecules encode instructions to synthesize a particular protein from its component amino acids ("translation"). Proteins have a wide variety of functions—they are the "workhorse" molecules of cells—and each has its own special role because of its unique structure, specified by the DNA and RNA sequences. Other DNA sequences do not specify an RNA sequence themselves, but can increase or decrease production of RNA from other DNA sequences.

It is increasingly easy and inexpensive to find out the complete DNA sequence of a person (or other animal). We can also find out the sequences of all the RNA molecules a cell has produced. We can often find out the sequence of amino acids that makes up a particular protein as well. However, in most cases we do not understand what the protein does or how it does it, which are much more challenging problems. In that sense, knowing a DNA sequence does not mean that we have "decoded" the DNA, because we cannot figure out the protein's function(s) just by looking at its DNA sequence.

In cases that get media attention, a correlation between a particular human DNA sequence and the occurrence of a particular behavioral variant is statistically strong enough that it is unlikely to have occurred by chance. In such cases, we might conclude that the DNA variant causes a particular form of the behavior. But what would we mean by saying that this DNA causes this behavior?

A variation in DNA may lead to synthesis of a slightly different protein, which may then function differently in a cell. A variation in the DNA that is not part of any gene may also have an important effect, not by changing any protein that is synthesized, but by changing how much of a particular protein is synthesized, as well as in which cells it is synthesized and when. But how does altering the amount, function, or location of a protein alter a behavior?

Beginning in the 1980s, some human geneticists attempted to find statistical correlations between individual variants of human genes and complex behavioral traits or conditions. Some christened this approach "one gene–one disease" (it was actually referred to as "OGOD") or one gene–one behavior (which we might call "OGOB"). Some of this research described "genes for" manic depression, schizophrenia, alcoholism, and homosexuality, among other complex conditions or traits.[2] Many of these findings could not be replicated, however, and may have been grounded on unjustifiable statistical assumptions.[3]

In retrospect, part of the problem may be that effects of genes on behaviors are usually very indirect (not proximate), and there is not a one-to-one correlation.[4] A gene that has an important effect on any behavior would act via the nervous system, which is what directly causes all behaviors. The proteins produced by some genes that have large effects on behaviors are key players in constructing the nervous system during development, particularly during prenatal development.[5] Many of these proteins act as signals (or allow neurons to respond to signals) that guide neurons either in their growth or in their formation of synapses connecting them to other kinds of neurons. These signals can also determine what sort of cell a particular neuron becomes and where it ends up in the nervous system, among other things. These are all extremely important effects. In fact, a small change in a key protein often has such momentous effects that the individual dies early in development.

But what such changes in a protein rarely do is alter just one behavior. Instead, a change in a protein that plays an important role in development, for example, typically changes a rule in how the "developmental game" is played. Alteration of a developmental gene may result in numerous aberrant connections between brain structures. Because each brain structure contributes to a wide variety of behaviors, the effect is not limited to one behavior. Since the developmental game is largely played out prenatally—before we observe the behaviors at all—and developing nervous systems may then compensate for alterations in the developmental game, it can be difficult to trace all the behavioral effects of a gene variant.

So the relationship between gene variants and behaviors is not one-to-one; it is one-to-many, because each gene with a developmental effect alters many behaviors. It is also typically many-to-one, because variants of many genes can have effects on each behavior.

More recently, most human geneticists have come to the conclusion that complex behaviors and common psychiatric disorders are affected by hundreds of genes, each of which has a very small effect on its own.[6] Many current studies are trying to distinguish the small contribution of each of many individual genes via genome-wide association studies.[7]

Individual gene variants that have a dramatically negative effect on any aspect of health are typically rare because they are selected against in evolution, as individuals with such genes do not have healthy children at a high rate. Huntington's disease—a relatively common fatal single-gene disorder—is an exception to this rule because most affected individuals do not show symptoms until after their child-rearing years.

Some reliable correlations between DNA variants and human behavioral variants have been and will continue to be discovered. However, in addition to the links between DNA variants and behaviors being one-to-many and many-to-one, such correlations reveal relatively indirect (not proximate) effects on behavior.[8] Many events occur in between a change in DNA and a change in behavior. And everything that happens in your life that changes the way you think, feel, and act has its effect by changing your nervous system. Many writers (usually nonscientists) have questioned whether there is a "biological cause" of normal or abnormal human behaviors, thoughts, or emotions. There is *always* a biological cause, because all of these things occur due to electrical signals in the nervous system. You simply cannot act, think, or feel differently without your brain generating different electrical signals in its neuronal circuits. The intended question is likely whether the cause is *genetic,* which is very different from asking if it's *biological.* The nervous system you are currently using to interact with the world is not just the result of your genes and those prenatal developmental events; it is also the result of countless other events that have occurred since then, including events that occurred today. Each of those events changed your nervous system. All these historical effects are mixed with and have interacted with the effects of your particular DNA variants.

Besides genes coding for proteins that help orchestrate the developmental game, there are genes coding for proteins that play key roles in electrical signaling within neurons or between neurons throughout our lives. Variants of these genes may alter, for example, the way that neurons signal to other neurons at synapses.

Some synapses are electrical synapses, where ions (charged atoms) flow directly from one neuron to another through a kind of protein tunnel (see Figure 2.1A). Others are chemical synapses, where the first neuron secretes a chemical whenever it generates a spike, and then this chemical (called a neurotransmitter) attaches to a protein (called a receptor) on the second neuron (see Figure 2.1B). Attachment to the receptor changes the receptor's shape, which in turn causes an electrical signal in the second neuron. A change to the neurotransmitter, its receptor, or how long the effect lasts (before the neurotransmitter is chemically broken down or carried away for recycling by neighboring cells) can have an important influence on whether the second neuron generates its own spike. Each type of neurotransmitter and receptor is typically deployed in many different parts of the nervous system, so any alteration in its signaling usually will affect several parts of the nervous system and many behaviors. So the relationships between variants of genes for neurotransmitters or receptors and behaviors are typically one-to-many and many-to-one, just as for genes that have developmental effects.

Although most individual genes have relatively weak or nonspecific effects on individual behaviors, a few have relatively strong and specific effects. Let's take a closer look at a couple of examples at this end of the spectrum.

The fruit fly *(Drosophila melanogaster)* has been an extremely useful species, especially for studying genetics (originally because of their large chromosomes, easy maintenance, and fast breeding, but now also because of the large number of methods that have been developed to manipulate them genetically). Like many animals, fruit flies show sex-specific behaviors used in courting and mating. One gene, called *fruitless,* comes in male-specific and female-specific forms (or, actually, its RNA and protein come in different forms, because the RNA is modified differently through splicing).[9] Altering the *fruitless* RNA in males dramatically alters a set of male-typical courting and mating behaviors. For example, altered males court both females and males, which unaltered males don't do. Female flies with the male

A. Electrical Synapse

1. A spike travels from the cell body along the axon.

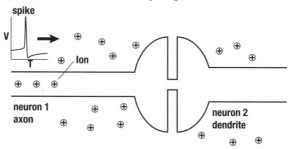

2. The spike reaches the synapse.

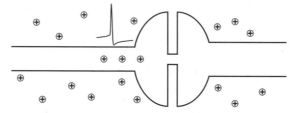

3. Ions go through a tunnel from the first cell to the second cell and change its voltage.

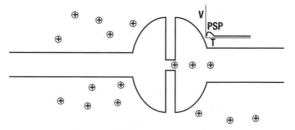

Figure 2.1. How a spike in one neuron causes an electrical signal (a postsynaptic potential, or PSP) in a second neuron at (*A*) an electrical synapse and (*B*) a chemical synapse.

B. Chemical Synapse

1. A spike travels from the cell body along the axon.

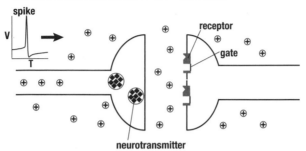

2. The spike reaches the synapse, causing neurotransmitter secretion.

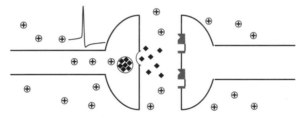

3. Neurotransmitters bind to receptors in the second neuron, causing ion channel gates to open and ions to flow.

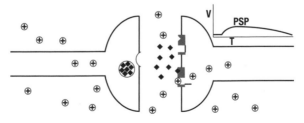

version of the *fruitless* protein show some male-specific behaviors, though not all.

How does the *fruitless* protein specifically affect this set of sexual behaviors? It turns out that the male version of the *fruitless* protein acts during development to guide the construction of neuronal circuits that are used in these behaviors, by altering transcription of other genes. While many of the details have not yet been worked out, it is now apparent that sex differences in development of neuronal circuits linked to the *fruitless* protein include some different synaptic connections that have important effects on sexual behaviors. So there are some cases in which individual genes (or at least their RNA) have strong and specific effects on a set of related behaviors. But even in these cases, the proximate control of the behaviors occurs through neuronal circuits. Working out how those circuits generate male- and female-specific sexual behaviors, and how the different forms of the protein alter those circuits during development, will be critical tasks for future studies of fruit fly courting and mating.

For another example, let's turn to mammals. Certain hormones (chemical signals that travel through the blood), and the receptors they bind to, can affect a whole suite of related behaviors in animals generally, including mammals. Often, the functionally related set of behaviors involves courting and mating, just as we saw for fruit flies. For example, oxytocin and vasopressin are hormones that alter reproductive behaviors, social bonding to a mate, and parenting behaviors in some mammals. This has been particularly well studied in the prairie vole, a small mouse-like rodent that forms a strong, lifelong bond to one member of the opposite sex (often referred to as monogamy) after mating, in contrast to closely related vole species that do not form such pair bonds.[10] It turns out that blocking the vasopressin receptor in males or the oxytocin receptor in females can prevent pair-bond formation.[11] Also, putting the prairie vole version of the vasopressin receptor into a part of the brain of a normally nonmonogamous vole species can lead to pair-bond formation.

In this example, much as for the *fruitless* gene in fruit flies, a single-gene change can lead to changes in a set of related behaviors. As with the *fruitless* gene, though, it will also be important in future studies to work out how the relevant neuronal circuits are altered and how they act to generate specific behaviors like pair bonding.

Besides these exceptional genes that have a large effect by modifying a suite of related behaviors, there are other gene variants, such as the one that causes Huntington's disease and those that increase the probability of getting Parkinson's disease or Alzheimer's disease, that act by killing brain neurons.[12] Killing neurons, of course, does have major effects on behavior, but each of these diseases kills neurons in multiple brain areas, affecting many behaviors.

So the bottom line is, gene variants can certainly affect behaviors, but the relationship between them is generally not one-to-one, and their effects are very indirect.

Brain Scans and Behaviors

What about using brain scans to find out how the human brain causes behaviors? Brain scan images show physiological changes inside a living human brain, with one or a few locations brightly lit up against a dark background, while the subject performs a particular task. These images are usually produced by either positron emission tomography (PET) or functional magnetic resonance imaging (fMRI). These kinds of experiments are increasingly common. For example, a search for scientific journal articles with "fMRI" in the article title or abstract in the database MEDLINE in July 2015 found 27,512 articles.

Brain scans have made possible valuable experiments that are in some cases not feasible using any other method. Brain scans can identify parts of the brain that are activated in response to specific kinds of stimuli, that are engaged in specific tasks, or that vary in their activation in correlation with behavioral differences. The exact functions of these brain areas can then sometimes be studied in more detail through other kinds of experiments, such as monitoring spikes in trained animals. Human brain scans also provide a means to study behaviors, cognitive functions, or abnormalities that are uncommon in nonhuman animal species or are unique to humans, such as self-awareness or schizophrenia. Comparisons between healthy individuals and those with a disorder can lead to hypotheses about the cause of the disorder. Images from brain scans can also engage public interest in neuroscience in a way that few, if any, other kinds of experimental data can.

Nonetheless, brain scan data also have limitations for understanding proximate mechanisms of behavior. The striking images from brain scans appear to provide a direct explanation of how the human brain performs each task. This perception can be misleading, however, for several reasons.

First, it is spikes in neurons that directly cause muscles to contract and thus cause behavior. Brain scans don't detect spikes in neurons. What both PET and fMRI measure are changes in blood flow or blood oxygen content.[13] What do blood flow or blood oxygen content have to do with anything?

It turns out that throughout the body (not just in the brain), tissues that "work harder" (i.e., metabolize at a higher rate) require more glucose and oxygen so they can make more ATP (adenosine triphosphate), the energy currency of our bodies. When tissues increase their metabolic rate, they are rewarded with increased blood and oxygen. Your body essentially operates according to the Marxist principle, giving "to each according to his need." When neurons receive more synaptic inputs and generate more spikes (are more "active"), they need more energy, and this leads to their receiving more blood and oxygen supplies. But the delivery of more supplies occurs after the need is already clear, that is, after the neurons have already been active for some time. So there is a delay between increased spikes in neurons and the increased blood or oxygen detected in the brain scan.

But how long of a delay are we talking about? It's only a few seconds. Does such a short period of time matter? Well, yes, it really does. A neuron's spike lasts for about 1 millisecond, which is one one-thousandth of a second. If this neuron sends a signal to another neuron, there will be a delay of perhaps 10 milliseconds before the second neuron generates its own spike. The second neuron might cause a third neuron to generate a spike maybe 10 milliseconds after that, and so on. So a lot happens in a neuronal circuit in less than a second. In fact, the sequence of events in a neuronal circuit that results in a movement is typically completed in less than a second. A brain scan detects a change in metabolism that occurs after this whole set of events has ended or while a person repeats a task many times.

This means that when a region of the brain "lights up" on a brain scan image during a behavior, we can say that this region had an increase in metabolism, but we can't say which part of the relevant sequence of neuronal events (i.e., which part of the circuit) that region contributed to. If the entire neuronal circuit were within that one region, this might not matter. But for most examples that have been worked out sufficiently, that doesn't appear to be the case. Instead, numerous and sometimes distant regions of the brain interact, often sending signals back and forth, during the planning and generation of each behavior. So it really is important to know which part of the neuronal circuit each region of the brain contributes to, and this kind of information cannot be obtained from a brain scan, at least by current methods.

But could we say that the neurons in this region contribute to all the behaviors that activate this region? Not necessarily. The reason is that each spot that lights up on a typical brain scan contains not one or a few neurons, but at least 100,000. (The smallest observation window—the spatial resolution—of the kind of fMRI used routinely, for example, is a voxel—a three-dimensional pixel—of about 1 mm^3, which contains at least 100,000 neurons.)[14] While this is a small fraction of the neurons in the brain, it is still a lot of neurons.

If each neuron within a particular brain region did the same thing, then it might not matter that so many neurons are imaged together. But this is rarely if ever true. They typically do very different things. They use different neurotransmitters and make different kinds of synaptic connections to different target neurons. Some release a neurotransmitter that excites other neurons, increasing the probability that

these other neurons will produce a spike, while others release a neurotransmitter than inhibits other neurons so that
they are less likely to produce their own spike. Adjacent neurons may generate spikes during different behaviors.

Finally, let's not forget that the bright spot on a brain scan
is not an image of the neurons; it is an image of blood vessels
that are "in the neighborhood" of the active neurons. Even
if we could be much more precise about the region that is
lighting up (perhaps through a new method such as photo-
acoustic tomography, which uses light to trigger tiny sound
waves[15]), it would not tell us which neurons have increased
their spiking during a behavior.

There are methods (including older methods, actually),
such as those employed in electroencephalograms (EEGs),
that detect human brain electrical signals, rather than blood
flow changes, and provide much more precise measures of
the timing of electrical signals. The limitation of an EEG,
however, is that it provides very poor spatial resolution: we
do not have a good idea of where the electrical signals originate (even within millimeters), and, in any case, each electrical signal detected on an EEG is actually a sum or average
of an enormous number of spikes and/or synaptic signals. So
although an EEG provides us with a better idea of *when* the
electrical activity occurred, we have an even harder time
saying *which neurons* increased their activity. A newer
method, magnetoencephalography (MEG), also indicates
the precise time of summed electrical signals and does so
with spatial resolution that is better than an EEG (though
not as good as fMRI).

Another newer method seeks to catalog the entire, enormous set of synaptic connections in a mammal's brain,
which together can be called the connectome (and so this

field is sometimes called connectomics).[16] These studies involve very sophisticated methods and can produce stunning images of brain anatomy. The ambitious goal of this field is essentially to create a map of all the brain's connections, which could then guide other kinds of experiments to help work out neuronal circuits and their functions.

It's important to emphasize, however, the limits to what neuroanatomy alone can deliver without monitoring electrical signals or behavioral outputs. For example, all the synaptic connections in the nervous system of the nematode worm, *Caenorhabditis elegans,* were cataloged decades ago using electron microscopy, yet this has not led to an understanding of which neuronal circuits generate which behaviors and how, despite the fact that the animal only has 302 neurons.[17] The problem is orders of magnitude larger for a mammalian brain with billions of neurons. So connectomics may provide a valuable anatomical database, but it will still be necessary to monitor electrical signals in neuronal circuits to figure out what they do and how they work.

Returning to DNA variants and brain scan data, if each provides only indirect links to behavior, rather than a causal sequence of events, why do we tend to focus on them as primary ways of explaining behavior biologically? One reason may be that we all gravitate to simple explanations that we can wrap our heads around and remember. There is a pleasing simplicity in these explanations, like a brief equation that explains how all particles move. Perhaps we'd like to think the control of behavior is simpler than it actually is.

There is a long history of research linking particular brain or head areas to particular behaviors, abilities, or abnormalities. Such research has sometimes provided a crucial first step in understanding the proximate cause of a behavior. In classical Greece, for example, Hippocrates and his followers noted that damage to one side of the human brain affects movements of the opposite side of the body.[18] In the nineteenth century, neurologists found that damage to a particular part of the human brain caused relatively specific deficits, such as in the ability to speak.[19] Such findings gave rise to the idea that specific functions are localized to specific parts of the brain, which was a conclusion that spurred further research into the roles these brain areas play in behaviors.

At times, however, the appealing simplicity of an equation between a brain area and an ability or abnormality has led researchers astray. This was apparent in the temporary success of phrenology, the attempt to link particular bumps on the skull with particular variations in human behavior. Phrenology was begun by Franz Joseph Gall in the late eighteenth century and was quite popular (especially in the United States and Britain) for the better part of a century.[20] In fact, a phrenology book, *Constitution of Man,* was at one time the third most common book in English-speaking homes. (The Bible was first.)

Eventually, phrenology was discredited and shown to have produced no meaningful data.[21] One problem was that the phrenologists relied on single-case studies, and another was that they focused on the skull (that's what was accessible to them), not the brain, even though the brain was what they were really interested in. Bumps on the skull don't necessarily indicate anything about the underlying brain. But phrenological studies were among the first to promote the idea that particular behaviors could be localized within the brain. This idea was further developed in the nineteenth century by neurologists studying human patients with brain injuries and in studies involving electrical stimulation of the brains of living animals, eventually including humans.

The use of brain scans, of course, is not phrenology. It is technically much more sophisticated. It appropriately focuses on the brain, not the skull. Conclusions from brain scan experiments are usually based on the study of multiple individuals performing each task, not just one person.

But interpretations of brain scan data sometimes have one thing in common with phrenology: the logic that we can explain a behavior or abnormality by identifying a location (or locations) in the brain associated with it. The point here is not that specific parts of the brain don't perform specific functions. They clearly do. The point is that such information is only the beginning of an explanation. A location in the brain does not by itself tell us how or why the behavior is produced.

Another reason for the appeal of brain scan data is the power of the images themselves. The bright, multicolor images are striking and aesthetically pleasing. The color scheme indicates the amount of increase in blood flow or blood oxygen level, so it carries information. But it also is eye-catching, attractive, and memorable. As a thought experiment, imagine what impression the same information would have on you if it were presented as a table of numbers or a series of graphs, rather than as an image of the brain. In an actual experiment, people rated a (fictional) cognitive neuroscience article as more credible when the article included brains scan images.[22] Perhaps for this reason, images of brain scans are common not only in the popular media, but also in textbooks of neuroscience, physiology, and psychology.

The way researchers typically display brain scan data has also evolved over decades to better draw attention to one or a few areas of the brain. For example, compare the early PET image in Figure 2.2A (left) to the more recent fMRI image in Figure 2.2B (right). The left image shows many gradations of blood flow, while the right image focuses attention on one brain area with the strongest increase in blood flow. The latter approach makes it easier for a viewer to see the main increase in blood flow, but arguably might mislead the viewer into an oversimplified view of how the brain produces behaviors.

The important point here is that finding locations in the nervous system that are linked to particular behaviors is just a first step in figuring out how the nervous system controls those behaviors. This point was made in 1930 by Karl Lashley, then president of the American Psychological Association (though he actually did experiments on the brains of a variety of animals, so he would be considered a neuroscientist in today's terminology), who wrote that the use of brain descriptions in psychology textbooks of the time "seems to provide an excuse for pictures in an otherwise dry and monotonous text."[23]

He added,

> Specialization of functions in the cerebral cortex is an indisputable fact, but we have yet to find an adequate interpretation of it. We have asked, Where are psychological functions localized in the brain? and have gained a meaningless answer. We should ask, How do specialized areas produce the details of behavior with which they

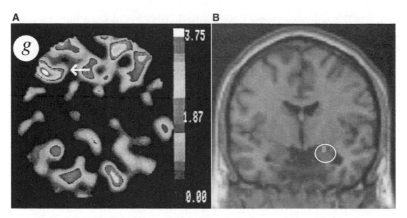

Figure 2.2. Examples of the evolution of human brain scan images. *A* shows a PET image from 1988 and *B* shows an fMRI image from 2002. See color insert.

are associated: what are the functional relationships between the different parts and how are they maintained?[24]

Lashley was *not* arguing here against the idea that particular brain regions have particular functions. Instead, he was appealing for experiments that would go beyond brain locations and address *how* behavior is controlled. In essence, Lashley was asking for an explanation in terms of the sequence of electrical signals in the neuronal circuit that controls a behavior. In other words, he was asking for the proximate causes of behaviors.

In the decades since then, techniques have been developed to monitor the electrical signals in individual neurons of neuronal circuits while animals produce behavioral outputs. As a result, we now know quite a bit about the complex relationships between the signals in neuronal circuits and the production of some behaviors, especially for species that are more easily investigated.

For the most part, we can't do this type of experiment (not yet, anyway) in people. There are obvious ethical issues with inserting electrodes into people's brains to monitor spikes in their neurons. (But this is actually done in exceptional cases, such as when a patient's brain is exposed for epilepsy surgery and they have volunteered for such a study, or when a paralyzed person receives an implant to control a prosthetic limb; we will return to these cases later.) But we can do experiments like this in a variety of nonhuman animals. When we do, we are able to reveal the chain of events through which electrical signals in neuronal circuits directly cause behaviors. Such studies reveal what kinds of neuronal circuits are responsible for each kind of behavior, and how.

Another reason it may be preferable to study the governance of behavior in nonhuman animals is that the human brain is larger and more complex than other brains, which makes it much harder to figure out what's going on. Of course, if you want to study something like how the brain produces religious thought, you might have no choice but to examine humans (and do brain scans, for example). But we share most of our behaviors with many other animals, and these behaviors can be studied in them to work out the neuronal circuits typically involved.

3

Neuronal Dictatorships

Command Neurons

Do all neurons in a neuronal circuit contribute equally to producing a behavior, or are some neurons "more equal than others"? Is a neuronal circuit something like a dictatorship, in which one neuron "calls the shots" and the rest of the neurons simply follow orders, or is it more like a democracy?

This question came into sharp focus for the neuronal control of animal behaviors in 1978, when Irving Kupfermann and Klaudiusz Weiss published an article entitled "The Command Neuron Concept."[1] This article appeared in a scientific journal *(The Behavioral and Brain Sciences)* that invites authors to write a piece (actually called a target article) that usually advances a provocative viewpoint. The journal then invites other scientists in the same field to shoot at the target, by writing commentaries that follow the target article. The Command Neuron Concept triggered a plethora of diverse commentaries. Basically, Kupfermann and Weiss had hit on something controversial.

Kupfermann and Weiss pointed out that the term "command neuron" had been in use for decades without really being defined. It was first used to describe neurons in crayfish that, when individually shocked electrically (which caused spikes in them), triggered coordinated movements of the crayfish's swimmerets (which look like little legs).[2] Since each swimmeret movement so produced looked complex and natural and was triggered by stimulating a single neuron, "command neuron" seemed like an appropriate descriptor for each of these cells.

But this name and its military analogy carry a heavy load of semantic baggage. They imply that this neuron sits at the top of a hierarchy. One imagines that, like a military commander, the command neuron collects signals from many sensory neurons. Then the command neuron makes a decision. If the command neuron decides to order a behavior, it sends a signal (a spike) that eventually causes a large number of peon neurons to make muscles contract. The highly organized chain of command, which might have many levels, ensures that the commander's order is carried out in a coordinated fashion, even if each peon neuron has no direct communication with the other peons.

But is this really what happens in a neuronal circuit? And can we tell what really happens just by shocking a neuron and seeing a movement in response?

Kupfermann and Weiss argued that people were using "command neuron" in a cavalier manner and needed to apply a rigorous definition instead. They suggested that to earn the name, (1) the neuron must normally spike just before a behavior starts, (2) stimulating the neuron must evoke the entire (normal-looking) behavior—that is, the neuron must be "sufficient" for the behavior, and (3) preventing the neuron from spiking must prevent the behavior—that is, the neuron must be necessary for the behavior.

From their survey of the scientific literature in 1978, Kupfermann and Weiss concluded that it was not yet clear that any neuron had passed all of these tests. Two of the best candidates, however, were neurons named the lateral giant and the medial giant in crayfish and a neuron called the

Mauthner cell (or M cell) in fish and tadpoles. Each of these neurons, or at least its axon, is huge. They are the biggest axons in that animal's nervous system (as is the squid's giant axon).

So do nervous systems arrange for the biggest neuron to rule them, much as the ancient Israelites apparently chose Saul as their first king because he was the tallest one? (Modern parallels might include how U.S. high school students often choose a large athlete to be their student body president or how U.S. citizens often elect the tallest major-party candidate to be their president.) Why would you want a big neuron to be your commander, anyway?

It turns out that bigger neurons (or at least, neurons with thicker axons) can send their spikes much faster toward the next neuron. So it might make sense to have a command neuron be big, to help it get its orders followed quickly. But when would speed of signal transmission be so important?

One answer is, when you need to escape from a predator. The giant neurons in crayfish and the M cell in fish and tadpoles each trigger a fast escape movement.[3] Crayfish flip (using that huge abdominal or "tail" muscle that we love to eat; see Figure 3.1), while fish turn away from the predator and then swim (see Figure 3.2). In each case, a single spike in a giant neuron is enough to trigger the entire behavior. And the behavior starts within about 10 milliseconds (one one-hundredth of a second) of the giant neuron spike. You can't get behavior much faster than that!

So these are genuine command neurons, right?

Well, it depends on whom you ask. Let's look at the lateral giant neuron and the M cell, using Kupfermann and Weiss's rigorous criteria.

Are these neurons normally active when the behavior occurs? Yes—it's been shown that a spike occurs in each giant

Figure 3.1. Sequential high-speed video frames show a lateral giant neuron–mediated tailflip in a juvenile crayfish (view frames from top to bottom). The abdomen was touched with a probe in the second photo. These photos were taken 10 milliseconds apart, so a total of only 50 milliseconds (one-twentieth of a second) elapsed from the second photo to the bottom photo.

Figure 3.2. Sequential high-speed camera frames show how a larval zebrafish turns away from a touch to the head in about 10 milliseconds (one-hundredth of a second).

neuron just before the behavior starts.[4] (You might wonder how it's possible to detect these tiny electrical signals while an intact animal is moving around in an aquarium. Remarkably, the giant neuron spikes are so large compared to the spikes of smaller neurons that you can detect them with a metal electrode just inside the animal, even while it's moving.)

If you shock the giant neuron while the animal is not really doing anything, does the animal escape? Yes—so the giant neuron is sufficient.[5]

If you prevent the giant neuron from spiking, does this prevent a normal escape from occurring? Well, it depends. If you stop the neuron from spiking by poking a glass microelectrode into the neuron and injecting negative ions into the neuron to inhibit it, which was Kupfermann and Weiss's recommendation, then a stimulus (like touching the animal) that normally triggers an escape response no longer does.[6] Well, that should do it, then; it's a command neuron by Kupfermann and Weiss's criteria.

But Robert Eaton and colleagues decided not to stop there. They used a second method to shut up the M cell: they heated it with a metal electrode until it was basically destroyed.[7] Then they put the fish back into the aquarium and dropped a ball into the water. And guess what? The fish escaped. But when they looked carefully, they found that each fish started its escape a bit later, about 37 milliseconds after the ball dropped instead of about 22 milliseconds. In a later study, the difference in escape onset with and without the M cell was just 4 milliseconds.[8]

Apparently, there are other neurons that can take over for the M cell when it is destroyed, but don't take over when it's merely silenced by current injection. Near the M cell, there is a group of smaller neurons, each of which makes similar synaptic connections. Essentially, you start with a giant dictator neuron in charge of escape movements, but if the dictator is assassinated, as Eaton and colleagues did, then an oligarchy of smaller neurons takes control of the escape movements, almost immediately. (This set of smaller neurons also appears to contribute to the later phases of the normal escape behavior, fine-tuning its trajectory.)[9] So we might still want to call the M cell a command neuron, but it fails the necessity test if you destroy it rather than inhibit it.

Still, you could argue that a fish without an M cell does *not* still exhibit the *same kind* of escape behavior; it exhibits a slower and less stereotypical kind instead.

But does it really matter if you start your escape 4–15 milliseconds late? Well, it might. It might make the difference between being eaten and not being eaten, which would be significant for most of us.

It turns out that many cases of apparent command neurons discovered so far do trigger an escape movement. Perhaps this should not be surprising. It's like a push button. If you're in a plane that's going down, you can hit the push button and eject. It's simple and fast and requires little deliberation.

Remarkably, the story with crayfish turned out to be essentially the same as the story with fish.[10] Crayfish normally use *both* giant and nongiant neurons for different kinds of escape flips.[11] Without the giant neurons spiking, crayfish can do a similar escape flip, but it's slower than when they use a command neuron. In an emergency, they rely on a giant neuron, but if they are aware of imminent danger, they use a nongiant neuronal circuit instead. A giant-neuron circuit produces a very stereotypical flip, while the nongiant circuits produce a flexible set of slightly different flips, through which movement direction can be better controlled. So if you have some time to collect information and deliberate, you might be better off using a nongiant circuit to make sure you're going off in the best direction, rather than using the eject button.

The neuronal circuit for crayfish escape involving the lateral giant (LG) neuron has essentially been worked out in its entirety.[12] The "circuit diagram" is illustrated in Figure 3.3. This understanding was made possible by the fact that a

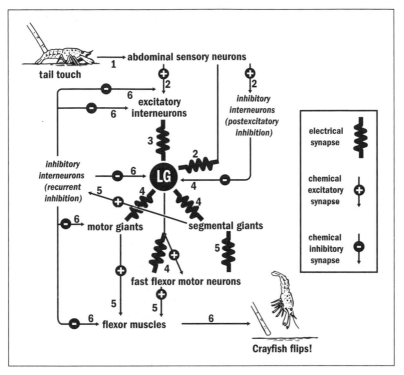

Figure 3.3. Diagram of the crayfish neuronal circuit responsible for escape movements via the lateral giant (LG) neuron. For each synaptic connection, the type of synapse is indicated (see the key in the box). Excitation is shown in roman font, inhibition in *italics*.

crayfish's abdomen, by itself in a dish of circulating saline, remains healthy for hours. Even just the abdominal nervous system by itself (without the shell or the abdominal muscles) remains healthy, and the entire LG neuronal circuit functions normally (but you have to electrically stimulate the sensory neurons, as the animal's shell is gone). The "virtual escape" produced in this way can be monitored via the spikes in motor neurons, which are the neurons that normally connect to muscles (although the muscles were removed in this situation). This was a tremendous technical advantage for figuring out the LG neuronal circuit, a nice example of Krogh's Principle.

Of course, this does not mean that the crayfish escape circuit functions *identically* in an isolated abdominal nervous system in a dish and in the intact animal, but its basic features are preserved, which has allowed scientists to figure out how it fundamentally works. However, there are nuances of its function that are not apparent in the isolated abdominal nervous system and that require a more intact animal to uncover, as we will soon see.

This circuit diagram allows us to "connect the dots" to explain how the LG neuronal circuit produces an escape. (You can follow along in Figure 3.3.) The basic story is that sensory neurons spike when the animal's abdomen ("tail") is touched (step 1). The sensory neurons directly excite LG via electrical synapses (2). (This has a small but fast effect.) The sensory neurons also indirectly excite LG via excitatory, chemical synapses to neurons in the middle of the circuit, called interneurons (2), which themselves excite LG via electrical synapses (3). (This has a larger effect.) If the total excitation of LG is sufficient, LG generates a spike. An LG spike directly (4) and indirectly excites two kinds of motor neurons, which in turn directly excite flexor muscles (5), and the animal flips (6).

This neuronal circuit has several striking features. First, the giant neuron really is at the top of a hierarchy. Spikes from many sensory neurons responding to abdominal touch are funneled to LG. A single spike in LG then initiates spikes in many downstream neurons, through a chain of command, to cause coordinated muscle contractions. So LG appears to be a dictator, living up to its name.

But is it certain that LG is "making the decision" about whether to escape? Could it be that the decision is actually made *before* LG gets its input, and thus LG is just following orders itself?

Gene Olson and Frank Krasne did a clever experiment to answer this question.[13] They triggered an escape by shocking an abdominal sensory nerve (causing spikes in sensory neurons), which in turn triggered an LG spike. But then they precisely varied the size of this shock so that it was sometimes too weak (there were too few sensory neuron spikes) to trigger an escape. To understand what's going on here, you need to know that when a neuron spikes, this is an all-or-none decision. A stimulus has to be big enough to excite the neuron beyond its spike threshold. A smaller, "subthreshold" stimulus also excites the neuron, but not enough, in which case the neuron's total output (spike or no spike) is nothing. So a small increase in a neuron's input can make the difference between zero output and one spike, which in the case of LG would be the difference between no escape and an escape.

Olson and Krasne reasoned that a small increase in the size of the shock would at some point cause whichever neuron(s) is actually making the escape decision to suddenly produce a spike. If this deciding neuron(s) is *before* LG in the circuit, LG would suddenly receive a much larger input, because this prior neuron would have just spiked for the first time, and then LG would be "forced" into its own decision to spike—essentially, following orders. On the other hand, if LG really is the one making the decision, the input to

LG would increase only gradually, though the LG output (whether it spiked or not) would change suddenly when LG reached its own spike threshold. What they found was that the LG input changed only gradually up until LG suddenly spiked, triggering an escape. So LG really is the one making the decision.

A second striking feature of the LG escape circuit is that there are other very large ("giant") neurons (or axons, at least) within it, besides LG.[14] There are also multiple *electrical synapses* (not just chemical synapses) within the circuit. Electrical synapses allow much faster communication between neurons than chemical synapses do. Ions flow directly from one neuron to another at an electrical synapse, without having to wait for a chemical to slowly diffuse to the second neuron, attach to a receptor protein there, and change its shape. (Okay, we're still only talking about a few milliseconds for a chemical synapse, but for the electrical synapse it's less than a millisecond.) The large axons and the electrical synapses both make the LG circuit operate unusually fast, minimizing the time between an abdominal stimulus and the resulting escape.

Third, there are some connections going *backward* in the circuit—what are called feedback connections. The feedback is generally *inhibitory,* or negative. What is this negative feedback doing there?

Suppose that LG has just spiked and the crayfish begins its escape flip. This is likely to cause further activation of abdominal sensory neurons, because the sudden movement of the abdomen against the water (or the water against the abdomen, if you prefer) would activate them. A strong activation of those sensory neurons would then cause a second escape movement, which would lead to a third escape movement, and so on. So the crayfish would just keep flipping forever, never even finishing one flip before starting the next. Not a good plan.

One type of feedback inhibition (called recurrent inhibition) causes LG (as well as the neurons that excite LG) to be quiet (unable to spike) for a while following each LG spike, ensuring that the escape movement LG triggers will be completed without interference by additional escape triggering.[15] Any later stimulation of the sensory neurons will then be due to a genuine threat, not due to the crayfish's own movements through the water.

The fact that the crayfish LG circuit has been worked out so thoroughly also allows us to ask additional questions about how it is altered in different circumstances. For example, crayfish display a simple form of learning (yes, crayfish can learn) called habituation, in which repeated, harmless stimuli cause less and less of a reaction.[16]

Suppose our crayfish has its abdomen stimulated repeatedly by something that really doesn't matter, or at least doesn't necessitate an escape. Maybe some vegetation is rubbing against it, or maybe its sibling is constantly shoving it. Annoying, yes, but not worth hitting the eject button for. How is our crayfish to avoid triggering an escape?

One way is to reduce the effectiveness of synapses from sensory neurons (and sensory interneurons) to their targets, which is called synaptic depression. Repeated stimulation that is insufficient to trigger an LG escape indeed causes synaptic depression at these synapses.[17] The synapses are depressed because less neurotransmitter is released each time the neuron spikes. As a result, a stronger stimulation of the abdomen is required to generate synaptic inputs to LG that are large enough to trigger an escape.

But this introduces another problem. An LG-triggered tail flip will cause many abdominal sensory neurons to be activated as the animal flips. As we've seen, LG triggers recurrent inhibition that immediately follows each LG spike and prevents LG from spiking again soon afterward. But doesn't habituation—via depression of sensory neuron synapses— still occur during the flip? It turns out that it doesn't, because not only is LG itself inhibited just after an LG spike, but the abdominal sensory neuron axons are also inhibited at a location just before their synapses with the excitatory interneurons.[18] Because of this "presynaptic inhibition," the synapses between the sensory neurons and the excitatory interneurons don't become depressed during a tail flip.

Moreover, even if there is no giant neuron spike, it turns out that each stimulus to the abdomen causes LG to be excited, *then inhibited* immediately afterward.[19] If LG were only excited, and the abdominal sensory neurons spiked repeatedly (before synaptic depression takes effect), then the excitation in LG would keep adding up until it triggered an LG spike—the sibling repeatedly shoving would eventually cause the crayfish to flip out. (Kids, please don't try this at home.) But because of the inhibition following each excitation (called postexcitatory inhibition), this doesn't happen. The inhibition cancels out the excitation when several weak abdominal touches occur in succession, so the crayfish remains relatively calm and avoids escaping until a true emergency arises.

The LG circuit also allows us to explore how social interactions modify the operation of a neuronal circuit that controls behavior. It turns out that it is not a dog-eat-dog world; it is a crayfish-eat-crayfish world. Crayfish, like many animals, compete with other individuals of the same species for territories and mates. But for crayfish, this is serious business, in which the loser of a fight is often killed and eaten. So if you're likely to be the loser, it pays to escape at the right time. How do you do that?

One way appears to be through the effects of a chemical called serotonin. You may have heard of serotonin, because it has received a lot of press in recent years for its potential role in human depression (antidepressants like Prozac are SSRIs—selective serotonin reuptake inhibitors—which delay the chemical breakdown of serotonin so its effects last longer) and an apparent correlation between the type of serotonin receptor a person has and their likelihood of acting violently. But serotonin is actually used as a chemical signal in many neuronal circuits and in many kinds of animals. It can act as a neurotransmitter, secreted at synapses to excite other neurons, or it can act as a neuromodulator, altering the operation of a neuronal circuit that uses other neurotransmitters.

In crayfish, serotonin (among other things) alters the likelihood of an LG-triggered escape. But here's the catch: whether serotonin increases or decreases the likelihood of an LG escape depends upon the *social status* of the crayfish. (For crayfish, social status is not "single" or "in a relationship"; it is fight winner or fight loser.) In dominant crayfish (fight winners), serotonin makes an LG escape *more* likely; in subordinate crayfish (fight losers), serotonin makes an LG escape *less* likely.[20]

That's right. When serotonin is around, the *winner* of a fight is more likely than the loser to perform an *LG* escape. What's the point of that?

The speculation is that a fight loser has to be ready to escape from a predator (which might include another crayfish) at any time. And it should be ready to use a sophisticated, directional escape, so it can make sure it gets out of trouble. This requires using *nongiant* neuronal circuits.[21] A fight winner, on the other hand, doesn't need to worry about anything—other than an unexpected attack from a serious predator (perhaps a bass or a trout). For this purpose, the push-button method of a giant neuron may be preferable.

The effect of serotonin on LG escape probability happens—at least in part—via a direct effect on LG, the decision maker. In dominant animals, serotonin causes an increase in LG's excitatory responses to abdominal stimuli; in subordinate animals, it causes a decrease in LG responses.[22]

How does the same chemical cause an increase in the excitation of LG in one animal and a decrease in the excitation of LG in another animal, just depending on its social status? Serotonin (like most signaling molecules in the body) can attach to different kinds of receptors, which have different effects. In LG, serotonin causes one effect through one type of receptor and causes the opposite effect through another type of receptor. When a crayfish fights a number of battles with rivals (assuming it survives), it alters its proportions of these two types of serotonin receptors, which in turn modifies its serotonin-modulated LG escape probability to match its recent win-loss record.[23]

So command neurons are used to trigger fast escape movements and provide a decision maker that can be modulated by changing circumstances. Other cases of command neurons for escape movements have also been found, for example, in mollusks and leeches.[24] A recent study showed that fruit flies have two parallel escape circuits, one in which a single spike in a giant neuron triggers a fast escape and another that is slower but more flexible, just like in crayfish and fish.[25] But are command neurons *only* used for escape behaviors? Apparently not. For example, recent studies have found a command neuron for feeding in fruit flies, and one for stridulation ("singing" by scraping one wing over the other) in crickets.[26] There could be many more such command neurons that just haven't been discovered yet.

Why feeding and singing? Are those emergencies? Well, no, but they are very important. Feeding is obviously how the fly sustains itself, and singing is how a male cricket gets a mate, which is at least as important as feeding, evolutionarily speaking. Perhaps what all these behaviors have in common is that they can succeed if they are stereotypical—exactly the same each time. A command neuron provides a quick and easy way to trigger a simple and stereotypical behavior.

Pontifical, Cardinal, and Grandmother Cells

So are command neurons a good model for how neuronal circuits usually operate? Do single neurons generally determine how a person or other animal thinks, feels, or acts?

It turns out that this kind of question has been around for a long time. William James, sometimes considered the "father" of psychology (and brother of the writer, Henry James), speculated in his 1890 textbook, *The Principles of Psychology*, about whether there is a neuron at the top of a brain hierarchy:

> There is, however, among the cells one central or pontifical one to which *our* consciousness is attached. But the events of all the other cells physically influence this arch-cell; and through producing their joint effects on it, these other cells may be said to "combine."[27]

This "pontifical" cell sounds a lot like LG, except that what James was talking about was not a decision to escape, but instead "consciousness"—something much more grandiose. But in terms of how neuronal circuits are organized, the issue is essentially the same. Do nervous systems in fact rely on a single neuron at the top of a neuronal hierarchy to make decisions or to provide awareness or perception of the world, or do nervous systems instead rely on equal contributions from many neurons, with no one neuron in charge?

Charles Sherrington had a very different opinion from James. He wrote in 1941,

> We might imagine this principle pursued to cul-mination in the final supreme convergence on one ultimate pontifical nerve-cell. . . . It would secure integration by receiving all and dis-pensing all as unitary arbiter of a totalitarian State. . . . But . . . the nervous system does not integrate itself by centralization upon one pon-tifical cell. Rather it elaborates a million-fold democracy whose each unit is a cell.[28]

This debate continued into the 1950s, as methods were developed to monitor the spikes of an individual neuron while a live animal was seeing something. Horace Barlow studied neurons in a frog's retina. He found that some neu-rons spiked a lot when the frog saw a fly-like object, so he speculated that these neurons are "fly detectors."[29] Jerry Lettvin and colleagues expanded on this in a 1959 article, "What the frog's eye tells the frog's brain."[30] They didn't sug-gest that each frog's brain has just one fly detector. But they did suggest that there are relatively few fly detectors in a frog's brain and that each fly detector sits atop a neuronal hierarchy.

In a 1972 article, Barlow crystallized this intermediate form of neuronal control, arguing that

> at the higher levels, fewer and fewer cells are active, but each represents a more and more specific happening in the sensory environment . . . and the "pontifical cell" should be replaced by a number of "cardinal cells." Among the many cardinals only a few speak at once; each makes a complicated statement, but not, of course, as complicated as that of the pontiff if he were to express the whole of perception in one utterance.[31]

So fly detectors are cardinals (or the other way around?).

The hierarchical view of neuronal circuits as applied to perception acquired another, nonpolitical metaphor in the 1960s. This was the idea of the "grandmother cell," meaning a neuron that always spikes when and only when you see (or perhaps even hear) your grandmother, no matter which way she is facing, what she's wearing, whether she's smiling or frowning, what the lighting is like, and so on. One imagines that this grandmother neuron receives a complex combination of visual and auditory sensory inputs that collectively provide evidence that grandmother is present; if so, the grandmother neuron sits atop a hierarchy of neuronal circuits and effectively weighs the evidence for Grandma.

Although the concept of a grandmother cell has long been familiar to neurobiologists, the actual origin of this phrase was obscure until Charles Gross wrote a wonderful article, "Genealogy of the 'grandmother cell.'"[32] Gross explained that the term was first used in its current meaning by Jerry Lettvin in a course he taught at MIT. As documented in a 1991 letter from Lettvin to Barlow, Lettvin told his students a story of a fictional neurosurgeon, Akakhi Akakhievitch (it seems likely that this name was taken from Akaky Akakievich Bashmachkin, protagonist of the 1842 short story "The Overcoat" by Nikolai Gogol), who was called upon to operate on Alexander Portnoy, the fictional protagonist of Philip Roth's *Portnoy's Complaint*. Portnoy came to see Akakhievitch because Portnoy was obsessed with his mother. Akakhievitch found and destroyed each "mother neuron" in Portnoy's brain (there was apparently not one, but several thousand of them), curing Portnoy of his neuroses. Following this episode, Akakhievitch despaired of winning a Nobel Prize for studying mother neurons and switched to studying "grandmother cells."[33]

So are there any grandmother cells?

As you might imagine, that's not an easy question to answer for the human brain, so Gross and other cognitive neuroscientists turned to rhesus macaque monkeys, beginning in the late 1960s. They focused their attention on a region of the temporal lobe of the monkey's cerebral cortex that appeared to have something to do with processing images of faces, based on studies of human patients and animal experiments in which that area of the brain was damaged.

These neuroscientists found some neurons in this region that spike selectively when certain faces—not just any face—are shown to them.[34] So these might actually be grandmother cells. But it's impossible to know for sure, because there is a very limited set of visual stimuli that can be used in any given experiment. How can we say that the category of visual stimuli that these neurons respond to is really the face of an individual monkey or human, as opposed

keep the task simple, initially there were only two directions of movement for these linked dots, and they were opposite directions.) The monkeys were trained to detect and report the direction of movement of the linked dots (not verbally, but by making a certain eye movement), in exchange for a sip of fruit juice. Monkeys and people are very good at getting a sense of which way the dots are moving, even if the dots moving together (the "signal") are a small fraction of the distracting dot movements (the "noise"). When they turned down the signal so that it was just barely more than the noise (which makes the task very hard—I've tried it!), then the monkey only guessed right part of the time.

They monitored spikes of one neuron at a time in the monkey's brain while the monkey was doing this. Here's where it gets interesting. The rate of spiking of some of these neurons was *even more sensitive* to the coherent movement of a subset of the dots than the monkey's behavioral choice was.[39] In other words, these individual neurons performed better on the task than the monkey did.

In addition, the rate of spiking of some neurons predicted the monkey's subsequent behavioral response *even when there was no right answer.*[40] That is, on some trials, there was not a subset of dots that moved in one direction, but the monkey still had to play the game and guess a direction. The spiking of certain neurons was well correlated with the monkey's guess, suggesting that these neurons reflected the monkey's decision, even when that decision was arbitrary. So each such neuron indicated the monkey's *choice* (and perception?), *not* simply the characteristics of the visual stimulus.

In a sense, these neurons are perhaps grandmother cells for motion detection, rather than for grandmother detection.

Of course, there could be many such neurons, but if they are all associated with making the same decisions, they are likely linked together in a perceptual circuit or a decision-making circuit.

In a variation on this experiment, the researchers made the linked dots move in one of eight different directions while they also electrically shocked neurons (to trigger spikes in them) that normally spike during a different direction of dot movement.[41] Instead of reporting an intermediate direction of dot movement, monkeys usually reported that one or the other direction, corresponding to the two stimuli, had occurred. Presumably, whichever movement direction got the most spikes in the brain won. This is called a winner-take-all mechanism.

So in this case, although it is not a single, grandmother cell that creates the perception, it might be a relatively small set of linked grandmother cells that all say the same thing at the same time.

So the bottom line is that neuronal dictatorships do control at least some behaviors in some animals. Most confirmed neuronal dictators command escape behaviors. Nonetheless, there is typically a group of other neurons waiting to take over quickly if the neuronal dictator is injured.

In addition, high-level perceptions, such as those required to recognize individuals, appear to be generated not by a single neuron, but by a relatively small set of neurons that behave similarly and collectively cause the perception. These neurons do not issue movement commands, but, like command neurons, they may be at the top of a neuronal hierarchy.

4

Neuronal Democracies

So are neuronal circuits usually organized hierarchically, with dictator neurons at the top, or are the examples I've given so far just exceptional cases?

A large body of work suggests that for most behaviors, neuronal circuits actually operate more like democracies. Each of these cases, in which a sizable group of neurons indicates a stimulus or decides on a behavior, can be called population coding, because a substantial group of neurons encodes information in the nervous system.

The story of population coding goes back even earlier than Sherrington, at least to Thomas Young, who formed a hypothesis in 1802 for how we are able to see colors.[1] You might imagine we see colors by having a different type of light sensor for each possible color (or actually for each light wavelength, for the technical purists). Each such sensor would be like a grandmother cell, but for a wavelength instead of a grandmother.

But here's what Young reasoned:

> As it is almost impossible to conceive of each sensitive point of the retina to contain an infinite number of particles, each capable of vibrating in perfect unison with every possible undulation, it becomes necessary to suppose the number limited, for instance, to the three principle colours,

red, yellow, and blue . . . each of the particles is capable of being put in motion more or less forcibly by undulations differing less or more from a perfect unison.[2]

What Young presciently envisioned is that we have only three kinds of color sensors (called "cone" photoreceptors, or just cones) in our retinas. Each type of cone actually responds to a wide range of wavelengths. So as a candidate to be a grandmother cell, each cone is hopeless. If it were activated by your grandmother, it would also be activated by, say, any old person.

But even though each cone provides only a ballpark estimate of the light wavelength, the three types of cones together could (and do) collectively provide a reliable and precise measure of wavelength. The reason is that each type of cone is more activated by some wavelengths than others, in a predictable way, and so the combination of x amount of activation of the blue cones, y amount of activation of the green cones, and z amount of activation of the red cones could only occur for one particular wavelength (see Figure 4.1).

As Edgar Adrian (Lord Adrian) put it in 1928 (he was thinking about how sensory neurons on your skin tell you where you've been touched, not how sensory neurons in your retina tell you what color you're seeing, but the basic idea is the same),

The accuracy of localisation will naturally depend on the number of receptors of the same type in the area which contains the stimulated region. . . . If there is much overlapping in the areas supplied by the terminal branches of dif-

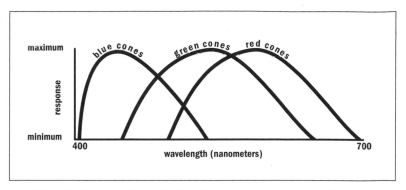

Figure 4.1. Diagram indicating the broad tuning to a range of wavelengths shown by each of the three types of cone photoreceptors.

ferent fibres it might be possible to localise the stimulus more exactly by comparing the relative intensities of discharge in the different fibres.[3]

In other words, not only is it possible to provide accurate sensing using sensory neurons that each respond to a wide range of stimuli, but you may actually be *more accurate* by using such imprecise neurons than by using precise neurons. The reason is that if each neuron responds to only a narrow range of stimuli, then any given stimulus will activate only a small number of sensory neurons. If one of those few sensory neurons is having a bad day and is not very accurate, then your perception of the stimulus could be way off. If, on the other hand, each sensory neuron responds to a wide range of stimuli, then *more* neurons will "vote" on the properties of each stimulus. If more neurons vote, and the nervous system calculates the sum or average of the votes, then the result may actually be a more accurate indication of the stimulus than a small number of precise sensory neurons would provide.

In fact, theoretical studies using data from computer simulations of neuronal circuits have suggested that if you have a certain number of neurons to use to indicate the properties of a stimulus, then these neurons will provide a *more* accurate answer if each neuron responds to a *wide* range of stimuli (or, in other words, is "broadly tuned").[4] This is because if more neurons vote, then their collective choice is more reliable. Another advantage in having many neurons contribute is that damage to some number of these neurons will have relatively less effect on their collective decision making.

In the case of human color vision, for each pixel you see, there are arguably only three voters—the three kinds of broadly tuned cones—but this voting process occurs at each location on the retina simultaneously.

It is now clear that this population coding occurs for all senses, not just vision, and also for the neuronal control of many types of movements.

One of the earliest examples was from the study of weakly electric fish (also called knifefish in pet stores) by Walter Heiligenberg and his colleagues. These fish generate their own electric field, which surrounds them. Changes in this electric field indicate the presence and the locations of prey, predators, competitors, and possible mates. (More on these fascinating animals in Chapter 7.) When another fish of the same species is nearby, they have to make their own electric field sufficiently different from their neighbor's field to continue sensing effectively, a behavior called the jamming avoidance response or JAR.[5]

Heiligenberg and his colleagues studied the neuronal circuit that produces the JAR, beginning in the 1970s. They noticed that neurons in the middle of the circuit are individually imprecise (broadly tuned) in sensing alterations in the electric field (unlike the JAR itself, which is very sensitive and reliable). As Heiligenberg later put it,

> The ultimate behavioral response of the animal would then depend upon the pooled input of such neurons. The JAR is thus controlled in an anonymous and parliamentary fashion. There is a marked absence of a few "pontifical" or "decision" neurons whose destruction could jeopardize the performance of the JAR.[6]

A particularly elegant example of population coding comes from crickets, which have air current–sensitive hairs on a paired rear structure called the cercus (see Figure 4.2). The cercal sensory neurons provide inputs to a set of four kinds of abdominal sensory interneurons that together encode air current direction, which tells the cricket which direction a predator is striking from. The response of each of these interneurons depends on the air current source direction in a way that looks almost like a sine wave (see Figure 4.3).[7] These graphs are unusually perfect for biological data! But each of these sensory neurons is still very imprecise, in the sense that it responds, to some degree, to air currents from a wide variety of directions. The key point is that the size of the response (the number of spikes) in each of these four types of neurons *in combination* yields a precise indication of air current direction, which then allows the cricket to escape in the best direction.[8] (In addition to providing coarse coding of general air current direction, recent evidence suggests that these interneurons also provide more precise information on local, dynamic components of

Figure 4.2. A paired structure on a cricket's abdomen, the cercus contains hairs that are each sensitive to air currents from a different direction. See color insert.

air current, which may tell the cricket whether it is a predator or something else that is causing the air currents.)

Similarly, leeches have a set of four broadly tuned sensory neurons to tell them where within each body segment they've been touched and therefore how to bend away from this touch. (This local bending behavior is shown in Figure 4.4.) These four neurons spike the most for touches to their top right, bottom right, top left, and bottom left. John Lewis and Bill Kristan were actually able to test whether the leech nervous system is using the information inherent in these sensory neuron signals.[9]

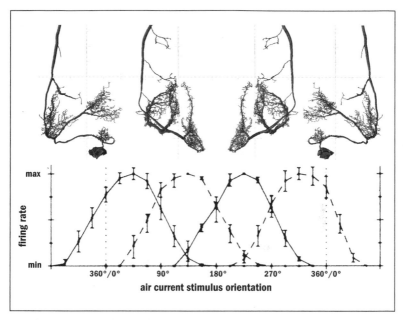

Figure 4.3. Rate of spikes in each of the four broadly tuned cricket abdominal sensory interneurons in response to air currents from different directions (with a tracing of each neuron above its tuning curve). See color insert.

To do so, they stimulated the leech neuronal circuit in two ways at the same time (see Figure 4.5). They touched the leech's body at one location (top right) while injecting positive current to trigger spikes in the sensory neuron that codes for another location (bottom right). The resulting bend was in a location approximately straight to the right. The direction of the bend shows that the nervous system used a compromise or average of the two stimuli. So we could say that the leech nervous system counts the votes from multiple neurons to determine its behavior.

It is not only invertebrates and fish that use population coding. Primates (like us) also do. Perhaps the earliest

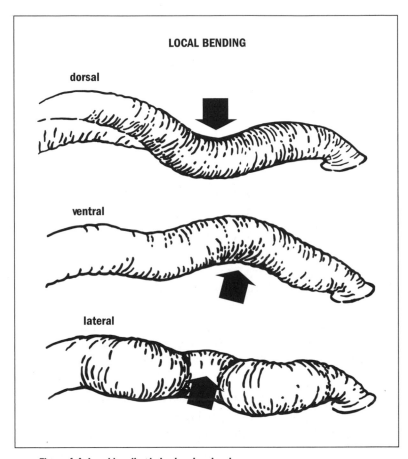

Figure 4.4. Local bending behaviors in a leech.

example of this was found in a part of the monkey's brain called the superior colliculus, which generates quick eye movements, called saccades, to visual targets. (You are making several saccades per second right now as your eyes jump from one word to another.)

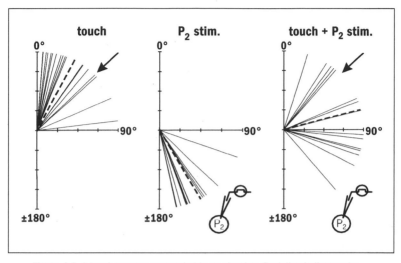

Figure 4.5. A leech nervous system holds an election. Each line indicates the direction of bending away from a stimulus during one trial. Left: When the top right part of the leech is touched (arrow), it usually bends away from this touch. Middle: When a particular sensory neuron (called P_2) is caused to spike, the leech bends away from the bottom right. Right: When both occur at the same time, the leech usually bends away from the right. This movement is approximately the average of the two votes.

In 1976, David Sparks and colleagues noticed that the neurons they were studying in the superior colliculus spiked when the monkey made certain visually guided saccades (the monkey had been trained to do this task), but were not very precise: they spiked for visual targets in a variety of locations, though each spiked most for one location and spiked less and less for locations farther away from this "preferred" location (see Figure 4.6).[10] How could such imprecise neurons be involved in producing a very precise behavior? Sparks and colleagues hypothesized that the entire set of neurons in the superior colliculus (probably millions of

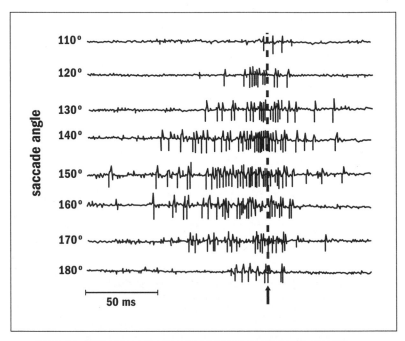

Figure 4.6. Spikes from one neuron in a monkey's superior colliculus were monitored while the monkey made saccadic eye movements (saccades) to a variety of locations. What look like vertical lines are spikes from this neuron. Each row shows spikes from this one neuron during a different saccade. The arrow at the bottom indicates the time each movement started. Note that this neuron spiked during many saccades but spiked most for the 150° saccade.

neurons) might contribute to each saccade through summation or averaging of their spiking.

Luckily for them (and us), the superior colliculus is organized in a way that made testing this hypothesis feasible: its neurons are arranged in a "map." Neighboring neurons in the map spike when the monkey makes saccades to neighboring locations in the world. Also, if you shock neighboring locations (triggering spikes) in the superior colliculus, you trigger saccades to neighboring locations in the world. The entire superior colliculus is systematically laid out like a grid, with one axis indicating how far to the side (left or right) a saccade will be and the perpendicular axis indicating how far up or down the saccade will be.[11]

Sparks and colleagues reasoned that if their hypothesis of population coding was correct, silencing all the neurons in one particular region of this map would alter most or all saccades in a predictable way. Basically, you would be holding an election in which a particular subset of voters is silenced. Choongkil Lee, William Rohrer, and Sparks actually did this, temporarily, by applying a local anesthetic to a part of the superior colliculus.[12]

The effect of the anesthetic was just what they predicted. For example, when the monkey tried to make a saccade to a light flash that was farther up and farther to the right than the saccade normally produced by shocking the now silent region, the monkey made a saccade that was *too far* up and *too far* to the right. The saccade was off target in exactly the way one would predict if neurons from the silenced region normally contribute to the behavior but could no longer contribute. This also means that a winner-take-all mechanism is *not* being used in this case. The population as a whole really does influence each saccade.

Also, saccades to the location normally targeted when the now silent region was shocked were *still on target.* This was possible presumably because the neurons surrounding the silenced region—those that normally spike for saccades that are farther up, farther down, farther to the left, and farther to the right—still contributed to saccade production and the *average* of their votes still yielded the *same* outcome as before the anesthetic was applied. So in this neuronal circuit, it is the average (not the sum) of votes that matters.

Population coding also appears to determine how primates reach with their arms. Neurons in a part of the cerebral cortex called the motor cortex spike for a variety of arm movement directions but spike most for a preferred direction, so they are broadly tuned to that direction.[13] It is probably not possible to do the kind of experiment with the motor cortex that Lee and colleagues did with the superior colliculus, because the motor cortex is not organized in a grid-like map, so we can't make a prediction about the effect of silencing neurons in a particular region. But Apostolos Georgopoulos and colleagues found that they could assess the population coding hypothesis mathematically.[14]

They created a formula that described how each motor cortical neuron changed its number of spikes, depending on the direction of the arm movement a monkey made (using actual data from experiments in which each neuron's spikes were monitored while trained monkeys moved their arms to various targets). Essentially, they replaced each neuron with a mathematical creature called a vector. A vector has a direction, or angle, and a length. The vector's angle was the movement direction each neuron preferred (the one it fired the most spikes for). The vector's length was the similarity of the target direction to the neuron's preferred direction—so if the target was in exactly the direction this neuron preferred, the vector would be longest, and if the target was in exactly the opposite direction, the vector would be shortest.

Then, they used vector addition to calculate the "population vector"—the sum of the votes of all the individual neurons—for each arm movement the monkey was instructed to make (see Figure 4.7). What they found was that this population vector quite reliably predicted the actual direction of reaching for each reach the monkey made (compare the

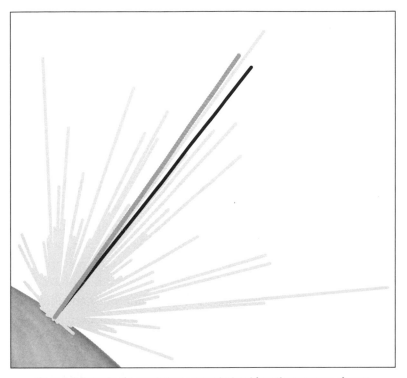

Figure 4.7. Example of a population vector calculated from the responses of many neurons in a monkey's motor cortex that spiked when the monkey reached in different directions. Each neuron's "vote" is indicated by a light gray line; the direction of the line is the arm movement direction it votes for; and the length of the line is how strongly it votes (its rate of spiking). The thicker and darker gray line indicates the direction of the population vector, which is the weighted average of all the light gray lines. The black line indicates the direction of the monkey's actual arm movement. Note that the population vector predicts the direction of the actual movement (the thick gray line is in almost the same direction as the black line), which suggests that the brain is holding an election to determine the movement outcome.

dark gray and black lines in Figure 4.7). In other words, the monkey really does appear to add together the signals from all these motor cortical neurons when it decides where to reach.

The few examples I've described are actually just the tip of the iceberg as far as the use of population coding to perceive the world and to control behaviors. Some of the other documented examples are the determination of escape direction (from touches or air puffs) in cockroaches; odor identity in insects; a fly's own motion while flying; sound frequency in fish; body location touched in turtles; sound source location in owls; several parameters of sounds in bats; location of an object touched by the whiskers in rats; tastes in mammals; somatosensory (touch) qualities in mammals; a seen object's orientation, length, movement direction, and speed in ferrets; leg position in cats; the amplitude and force of arm movements in monkeys; and the lengths of muscles in humans.[15]

The bottom line is that neuronal dictatorships do occur in some circuits, especially for dealing with emergencies and stereotypical movements, but population coding is much more common.

Spying on the Brain

If neurons are constantly spiking to determine which behavior will be produced, then in principle we could monitor this spiking to predict the decision. But why would you want to do that?

There are several classes of human patients that cannot voluntarily control their movements, particularly their arm and leg movements. For example, people with severe spinal cord injuries cannot voluntarily move parts of their body below the level of the injury. People who have had an arm or leg amputated no longer have that body part to move; they might get a movable prosthetic limb to replace it, but how could they voluntarily control it? Also, patients who have "locked-in syndrome" are unable to move at all (except sometimes their eyes), yet are fully conscious and aware.

In each of these cases, if you could monitor spiking in brain neurons that contribute to deciding on a voluntary movement of the arm or leg, you might be able to tally their votes and use the outcome (via a computer program) to trigger the corresponding movement of the natural or prosthetic limb. This arrangement is called a brain-machine interface (BMI) or brain-computer interface.[16]

How could this be done? You could noninvasively monitor electrical signals from the brain with an EEG and use these signals to control the limb or prosthesis. But as we've seen, these are very crude signals that represent averages of millions of neurons' voltages. It's not possible to figure out from these signals what precise movement a person is trying to produce.

Better information can be obtained by placing very narrow wires or needlelike microelectrodes into the brain itself. Engineers have designed "arrays" containing several rows of evenly spaced microelectrodes, through which you can monitor spikes from one hundred or more neurons (from just outside these neurons). You can place an array in a region known to play an important role in controlling movements, such as the motor cortex, and monitor spikes from many individual motor cortex neurons simultaneously.

How should you count the votes? Perhaps the simplest method would be to calculate the population vector. As we've seen, this calculation assumes that each neuron always votes for its own favorite movement direction, but the more spikes it generates at the moment, the more times it gets to vote. In other words, its contribution is weighted by its spike rate.

There are other ways of "decoding" the meaning inherent in the neurons' spiking, however, some quite complicated. We don't really know which decoding methods the nervous system itself uses to generate movements from these signals, so for the purposes of creating a useful BMI, we can use whichever method(s) works well.

How many neurons do you have to monitor to get a reliable measure of the intended movement? There are probably hundreds of thousands or even millions of neurons normally involved in producing a voluntary movement, so one might imagine you would need to sample quite a lot. Surprisingly, though, the motor cortex population vector is reliable within about 20 degrees of the intended arm-reaching direction when only about one hundred motor cortex neurons are monitored, in monkeys at least.[17]

One also might imagine that it's important to sample the "right" neurons. It would seem to be most important to sample neurons from the brain regions that normally contribute most to selecting or producing the movement, such as the motor cortex. This, indeed, is where most neuronal recordings for BMIs have been focused. Other areas of the brain, though, such as the parietal lobe (behind the motor cortex, which is in the frontal lobe), have also furnished useful signals.

From what we've seen so far, then, you should be able to get the required information by monitoring enough neurons in the right region and then decoding their spike code suitably. It also can help, it turns out, to send signals back to the brain from the natural or prosthetic limb to indicate the current position of the limb, mimicking sensory neuron feedback pathways.

Indeed, BMIs like this have been used successfully (depending on your measure of success), not only in monkeys, but also in human patients. In the earliest experiments, patients were simply able to move a cursor on screen using only their thoughts, but in more recent versions of BMIs, some patients (who could not otherwise perform voluntary limb movements) have been able to make more sophisticated and nuanced movements, such as picking up a cup and drinking from it.[18] We are still in the early days of this kind of research, and it is likely that BMIs will continue to get more sophisticated and allow fancier voluntary movements.

Remarkably, though, it might be possible to control a BMI suitably without monitoring a large number of neurons or the "right" neurons or using a suitable decoding method.

Chet Moritz, Steve Perlmutter, and Eberhard Fetz trained monkeys to play a video game (see Figure 4.8).[19] Each monkey used its wrist to control a computer cursor. A box appeared on-screen and the monkey had to move the cursor into the box to obtain a juice reward. This was a pretty simple game and the monkeys quickly learned it.

Then, the experimenters made the game harder. They monitored spikes from one to two neurons in the monkey's motor cortex (via microelectrodes) and used the spikes in these neurons, instead of the wrist movements, to control the cursor's movement. For example, if a neuron's spike rate increased, the cursor moved in one direction, and if its spike rate decreased, the cursor moved in the opposite direction. Once the monkey had learned this version of the game, the experimenters anesthetized the monkey's arm, so it could no

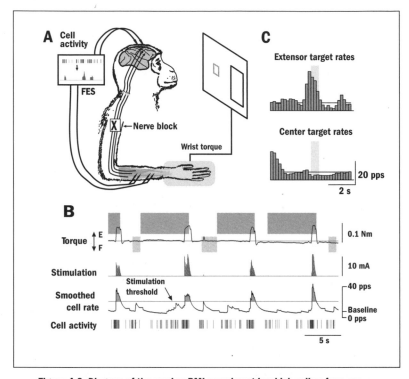

Figure 4.8. Diagram of the monkey BMI experiment in which spikes from one neuron in the motor cortex were used to control wrist extension. The dark gray rectangles at the top of *B* show when the target box was on the right side of the computer screen, requiring the monkey to extend its right wrist to move the computer cursor into this box to obtain a reward. The vertical lines in the cell activity row of *B* indicate the timing of the spikes from one neuron in the motor cortex. The smoothed cell rate in *B* indicates the rate of these spikes. Whenever the rate was above the "stimulation threshold" (gray horizontal line), the wrist extensor muscle was electrically stimulated and the wrist extended (torque), causing the cursor to move into the box. *C* shows that the spike rate went up when the on-screen box was to the side (extensor target) but not when the box was in the center (center target).

longer move its wrist in the usual way, and used the spike rate of the motor cortex neuron(s) to directly trigger electrical stimulation of a wrist muscle (bypassing the spinal cord and its motor neurons), which in turn moved the cursor. For example, an increased rate of spikes in one neuron would be used to electrically stimulate a wrist extensor muscle to contract, and the cursor would then move to the side and into the box. The monkeys were still able to play this version of the game successfully, because the spikes in the motor cortex neuron occurred at the right time to trigger electrical stimulation of the wrist muscle appropriately.

To succeed in this game, the monkey had to *alter* the spiking of this one neuron in its brain to make it spike at the right time and the right rate, following each on-screen instruction. You might imagine that this could only work if they happened to be monitoring just the right neuron to begin with—in our example, one that normally spikes just before the wrist extends. But that was *not* the case. In fact, it didn't matter in what situation that neuron had been spiking before the game began. Regardless of what the neuron had been doing previously, the monkey learned to make this one neuron spike at the right time and rate to move the cursor into the box—and did so within about thirty minutes of starting the game!

But is it really plausible that the monkey can target an individual neuron in its brain to change its spiking? How does it know which one to change, and how does it affect just one? Perhaps a more plausible explanation is that there are many neurons whose spiking is somehow linked together and the monkey is changing all of them at once. This would be analogous to the idea that there are grandmother neurons, but there are many such neurons for the same grandmother and they all spike at the same time. Recently, a human patient found he could alter the spiking of one neuron in his parietal lobe (monitored by an implanted electrode array and displayed to him on a screen) by imagining making a particular movement, such as moving his hand to his mouth.[20] Presumably, he activated a whole circuit of neurons that control this kind of movement, which included the monitored cell.

What does this mean for future use of BMIs? Well, it might mean that monkeys and people are so good at learning to change the spiking of an individual neuron in their brain, given adequate feedback on their progress, that they can learn to use a BMI successfully even if the *wrong* neurons are sampled and they are decoded in some arbitrary way. Maybe we don't even need to sample many neurons. In any case, these experiments indicate an extraordinary ability to modify neuronal circuits as needed. More on the modifiability of circuits in Chapter 6.

5

How Are the Factories Run?

Our most common behaviors are essentially the engines that keep our bodies going. They are our factories, producing not automobile parts or plastic toys but the movements that we depend on for life. Many of these fundamental movements occur cyclically and rhythmically. Some we must do almost all the time to stay alive: our hearts must beat, we must breathe, and we must digest.

There are other rhythmic behaviors that we do only some of the time—episodically—yet they are also critical parts of our behavioral repertoire. These episodic, rhythmic behaviors include walking and running (as well as swimming and flying, if you happen to be a fish or a bird, for example). These examples are all forms of locomotion, which we normally depend on to get from place to place (not counting southern Californians, who normally drive to the mailbox).

Rhythmic activities constitute a special group that requires special explanations. What is the origin of each rhythm, and why is each activity cyclical? Why are some rhythms continuous while others are episodic? What keeps these rhythms from stopping, which could be life threatening?

Anatomically, our rhythmic movements are produced in different ways, using the three different kinds of muscles in our bodies. We use cardiac muscles (our heart chambers) to pump our blood. We use skeletal muscles (the only ones we have voluntary control over) to breathe and to walk or run. And we use smooth muscles to break up food and move it along the digestive tract.

Yet in all these cases, the rhythmic activities are caused by rhythmic electrical signals. One could say that there is an electrical circuit that generates each rhythm. What are the components of each of these electrical circuits? Where are they located? How do they work together to generate a rhythm?

Is the neuronal circuit that causes each rhythmic behavior like a government program, housed within our central nervous system? Does it run autonomously, without feedback?

Or alternatively, is each rhythmic behavior a series of brief, separate actions? Is it like a free market economy, in which each action causes feedback (from sensory neurons) that triggers the next action?

The Heart of the Matter, or the Matter of the Heart

Let's start with the heart. It has been known for ages that hearts can beat even without input from nerves. In fact, a frog or turtle heart, taken out of the body and placed in a dish of saline, can continue beating for hours. This tells us that a vertebrate heart does not require a nervous system to make it contract rhythmically. It can generate the rhythm all by itself. The heartbeat is controlled by an autonomous program. In this case, however, a very local government—the heart itself—creates the rhythm. How does a heart generate a rhythm on its own?

The heart's rhythm normally depends on a special subset of heart cells, called pacemaker cells. Pacemaker cells do just what their name suggests: they generate the electrical rhythm of heart contraction, which is then relayed to other heart cells via electrical synapses. A person gets an artificial pacemaker implanted in their heart if their pacemaker cells are not doing their job adequately.

So one thing that's already clear from this is that it's not only neurons that make electrical signals—heart cells do, too. In fact, all kinds of muscle cells make electrical signals, although they can be different kinds of electrical signals.

But how does a heart pacemaker cell generate an electrical rhythm? It can't be through a neuronal circuit, because a heart cell is not a neuron and its rhythm is generated by a single cell, not by a group of cells.

In a sense, a pacemaker cell contains within it an entire electrical circuit. The components of this circuit are not individual cells, connected by synapses. Instead, they are individual types of ion channels, which are proteins in a cell's membrane that act as gatekeepers for certain kinds of ions to flow in and out of the cell. Ions are charged particles. In our bodies, the key types of ions for electrical signaling are sodium, potassium, calcium, and chloride. Sodium, potassium, and calcium are positively charged, and chloride is negatively charged.

Spikes occur in neurons (and also in most types of muscle cells) because of ion channels. Before each spike, the inside of a neuron (and essentially every other kind of cell in our body) has a slightly negative net charge compared to the outside of the cell. This is mostly because positively charged sodium ions are more concentrated outside the cell than in.

In this way, a cell is like a battery, maintaining a voltage difference between one side of the cell membrane and the other. In a battery, current flows when a wire connects the two terminals. In a cell, current flows across the cell membrane whenever ion channels in the cell membrane are open, allowing particular kinds of ions to flow across the membrane.

During a typical neuronal spike, sodium channels first open their gates. Sodium ions can then move freely through these open gates, into and out of the neuron. Sodium ions mostly rush into the neuron from outside, because sodium ions are more concentrated outside the neuron than in (and also because the positive sodium ions are attracted to the slightly negative inside of the neuron). This movement of positive sodium ions into the neuron generates the first half of the spike, when the inside of the neuron becomes positive (see Figure 1.1 in Chapter 1).

Then the sodium channels' gates close and the potassium channels open their gates. Potassium mainly rushes *out of* the neuron because potassium is more concentrated inside the neuron than out. This generates the second half of the spike, when the inside of the neuron goes back to being negative.

A heart pacemaker cell has types of ion channels similar to those in a typical neuron, but also has additional types. The effects of each type of ion channel on another type form a kind of electrical circuit within the cell that generates each cycle and so makes the heartbeat continue indefinitely (see Figure 5.1).

In a heart pacemaker cell, the positive ions that rush in to make the inside positive are mainly calcium ions, rather than sodium ions (and they come in through different kinds of ion channels). As with a neuron, however, potassium ions then rush out of the pacemaker cell, making it once again negative on the inside.

If that were the end of the story, a heart pacemaker cell would generate an electrical signal similar to a neuronal

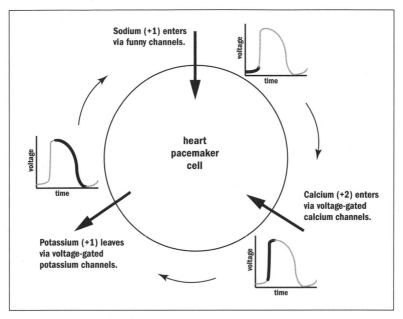

Figure 5.1. Diagram of a heart pacemaker cell, showing the continuing cycle of events as ions move in and out through voltage-gated ion channels. Bold curves show the voltage change that occurs in each step of the pacemaker cycle.

spike and would not be a pacemaker at all. But the pacemaker cell has an additional type of ion channel up its sleeve that does a special trick. This type of ion channel is very strange, so it was named a "funny channel."[1] What makes a funny channel funny?

The sodium and potassium channels that cause each neuronal spike open their gates when the inside of the cell becomes less negative. For example, if a few extra positive ions come into a cell (by whatever means), the sodium and potassium channels nearby will then open their gates. A change in the net charge (or the voltage) inside the cell causes them to open, so they are called voltage gated. The voltage-gated sodium and potassium channels that cause a neuronal spike both open their gates because the voltage inside the cell goes up (becomes less negative). But the sodium channel gates open (and close) faster than the potassium channel gates do, which is why the voltage first goes up and then goes back down during a neuronal spike.

A funny channel is funny because it opens its gates at a strange time. It is voltage gated, like the neuronal sodium and potassium channels we discussed, but it opens its gates when the voltage inside the cell goes *down,* rather than up. What difference does this make?

The voltage inside a cell goes down after the potassium channels have opened and the potassium ions have rushed out, bringing the cell back to its negative, "resting" voltage. Most neurons would remain "at rest" then unless something else happened to wake them up.

But that's not what happens in a heart pacemaker cell, because of the funny channels. When the voltage goes back down, that's what triggers the funny channels to open their gates. When their gates open, both sodium and potassium ions can pass through them. They tend to move in opposite directions, but sodium mostly goes through these gates, rushing into the cell, so the voltage inside the cell starts to go back up. Once it goes up enough, the voltage-gated calcium channels open (and the funny channels close) and the cell is on its way to another upward swing. Once the potassium channels open again and bring the voltage back down, the funny channels open again and send the voltage right back up. So the voltage goes up and down, up and down, like clockwork, for the life of a person (or other animal). This combination of ion channels effectively forms an electrical circuit that makes the pacemaker cell a pacemaker.

There is actually a second mechanism that also makes the pacemaker cell's voltage go up and down. Calcium ions are alternately released from a storage depot inside the cell and then brought back into storage. This "calcium clock" is normally synchronized with the previously described "membrane clock" involving funny channels. The two circuits work together to generate the rhythm, but the relative roles of the two are still debated.[2]

Of course, your heart does not always beat at the same rate. This is because both the membrane clock proteins and the calcium clock proteins in heart pacemaker cells have their functions altered by neurotransmitters from brain neurons and by hormones. (We'll learn more about what such neuromodulation can do in Chapter 6.) Each neurotransmitter or hormone either increases or decreases the pacemaker's pace. Norepinephrine and epinephrine (which is also called adrenalin) increase the pace, while acetylcholine decreases the pace.

In fact, the slowing of a frog's heart rate by acetylcholine was what allowed Otto Loewi to discover chemical synaptic transmission in 1921. Loewi electrically stimulated the vagus nerve to the heart of one frog to slow its rate, then removed some fluid from around this heart and squirted it onto a second frog heart, which slowed the second heart's rate. This showed that a chemical released by the vagus nerve (later shown to be acetylcholine) alters heart function.

Loewi literally dreamed up this experiment, two nights in a row.[3] As soon as he awoke from the first dream, he scribbled notes on a piece of paper, but he later couldn't understand what he had meant. After he awoke from the same dream the second time, he went to his lab and performed the experiment. His demonstration of chemical neurotransmission eventually led to the discovery of many other neurotransmitters throughout the nervous system.

Central Programs versus Feedback Mechanisms and the Roles of Rebellious Young Scientists

So what about the other factories of life? Are they also basically run by local, autonomous pacemaker cells? Or are they run by neuronal circuits within the central nervous system? Or is no one running the factories at all? Is the apparent coordination of movements actually the result of an ongoing series of separate events, each of which triggers the next, like a free market economy in which coordination emerges without a coordinator?

It turns out that any or all of these mechanisms can contribute to running the factory, depending on the factory. Rhythmic digestive movements in mammals do depend on pacemaker cells that are located within smooth muscle layers that line our gastrointestinal tracts (but their rhythms are generally much slower than the heart's). But the story for rhythmic behaviors produced by skeletal muscles, which require the nervous system, is more complicated. The governance of rhythmic skeletal muscle movements has been debated for generations and is still controversial. There is evidence that pacemaker neurons in the central nervous system play important roles, but there is also evidence that central neuronal circuits made up of nonpacemaker neurons connected by synapses contribute to most or all of these rhythms. Feedback inputs from sensory neurons also modify the rhythmic outputs. In most cases, we are still trying to figure out how all of these mechanisms work together.

Let's go back to the origin of one of these controversies. In the early twentieth century in England, Charles Sherrington was doing a variety of experiments on a variety of animals, but especially on the spinal cords of mammals.[4] It had been known for some time that most vertebrates can still make coordinated leg movements even when their brain is cut off from their spinal cord. In particular, they can still do locomotion (walking, in cats and dogs) and they can still scratch, which is also a rhythmic behavior. These experiments showed that the brain is not required for these behaviors. (You're probably familiar with the phrase "running like a chicken with its head cut off.") Removing the brain does not remove the entire central nervous system, however, because the spinal cord is also part of the central nervous system. But it still wasn't known how these rhythmic movements are generated.

Sherrington hypothesized that a chain of reflexes occurs, each one triggering the next. By "reflex" (this word gets used in a lot of different ways), Sherrington meant a movement that is triggered by immediately prior spikes in sensory neurons, which then (directly or indirectly) cause spikes in motor neurons. Each movement itself triggers spikes in some sensory neurons, which in turn trigger another movement. This chain of reflexes continues on and on, so, for example, a cat can keep moving its legs back and forth as it walks or runs.

In this hypothesis, each movement would occur in response to feedback from a previous movement, without a central coordinator. If the first movement were different, the subsequent movement would also be different. But what would make these movements occur in an appropriate sequence to generate an apparently coordinated behavior?

Sherrington imagined that when a leg moves in one direction, let's say forward (which involves contraction of leg flexor muscles), this movement would excite a specific group of sensory neurons (for example, the stretch receptors in the same leg's extensor muscles, which would be lengthened by the forward movement of the leg). These sensory neurons would generate spikes, which would in turn excite motor neurons and then muscles (through synapses) that would then contract to produce the *opposite* movement (in our example, extensor muscles would then contract and make the limb move backward).

Then, this second movement would excite a second group of sensory neurons (in our example, this could be stretch receptors in the leg's flexor muscles, which would be lengthened by the backward movement of the leg). The second group of sensory neurons would in turn excite motor neurons and muscles (flexor muscles, in our example) that would contract to regenerate the first leg movement (forward). These cycles would continue on and on—a chain of reflexes.

This was a great idea, but does it really happen? How could it be tested? One way would be to somehow remove the sensory neurons that the hypothesis requires and then see if the rhythm could continue anyway. But how could that be done?

Previously, it had been shown that a dog held above the ground in a harness could still make walking movements with its legs (called a "mark time" reflex). So the sensory neurons activated when each foot touches down are not necessary to generate the walking rhythm.

Also, when the nerves containing the axons of sensory neurons coming from the lower parts of the legs were cut or anesthetized, a dog could still walk. So none of these sensory neurons were necessary. But that still leaves the sensory neurons coming from the upper parts of the legs.

So Sherrington did a trickier experiment. He cut what are called the dorsal (or posterior) roots. The dorsal roots are essentially sensory nerves that contain all the sensory neuron axons from all types of sensory neurons (skin, muscle, joint, etc.) for each segment of the spinal cord. (The spinal cord is organized into anatomically repeating segments as one goes from neck to tail.) He surgically exposed these dorsal roots (they are right next to the spinal cord) and then cut all the dorsal roots for one hind leg of a cat or dog.

To test the effect of all these cuts, he triggered scratching, not walking.[5] Scratching is easier to trigger—he could just tickle the skin near the ear. He found that the cat or dog could still scratch just fine with all the hind leg dorsal roots cut. So it seems that sensory neurons from the leg are *not* required to produce cyclical movements, at least for scratching. But, oddly enough, Sherrington did not come to this conclusion. He did note that events within the spinal cord must be important to generate the scratching rhythm, but he still argued for his chain of reflexes hypothesis.

Arguably, Sherrington's hypothesis was still tenable because he did not cut the dorsal roots from the *other* hind leg, which might be excited rhythmically during each scratching movement (via vibrations from the moving limb that are transmitted through the body) and trigger the subsequent scratching movement. But Sherrington and his junior colleague, a young Scotsman named Thomas Graham Brown (usually known as Graham Brown), did another experiment in which they cut the dorsal roots to *both* hind legs.[6] These animals apparently still scratched, which would seem to rule out Sherrington's chain of reflexes hypothesis as a required mechanism, at least for scratching.

I say "apparently" because (as far as I know) the data from this experiment were never published in a scientific journal, which is how scientists (then and now) document their findings and expose them to evaluation by other scientists in their field. Sherrington and Brown did, however, actually bring such a cat to a scientific meeting to demonstrate their results in person (or in cat).[7]

So Sherrington appears to have been wedded to his chain of reflexes idea and reluctant to give it up even in the face of data that appeared to undermine it. But not Graham Brown. Brown did further experiments with cat walking, which he described in a 1911 article that he wrote alone (without Sherrington as a coauthor).[8] Brown cut all the dorsal roots for one hind leg of a cat and also cut almost all the nerves to that leg that contain the axons of motor neurons, so that the only intact neuronal pathways in that leg were the pathways from motor neurons (in the spinal cord) to two muscles, one a flexor muscle and the other an extensor muscle. (He also cut everything to and from the other hind leg.)

Then, Brown cut the spinal cord, because it had previously been found that this one-time event (probably because it was suddenly removing inhibitory inputs that had come from the brain) could trigger walking movements of the hind legs. When he did this in the cat with all the leg dorsal roots cut, the two leg muscles still connected to the spinal cord contracted rhythmically and alternated with each other, just as they do during walking (see Figure 5.2).

So, Brown concluded, sensory inputs from the leg are not required for the walking rhythm. Instead, the spinal cord is able to generate the walking rhythm all by itself. The output of the spinal cord is similar to the output of a pacemaker cell: it cyclically and rhythmically excites flexor and extensor

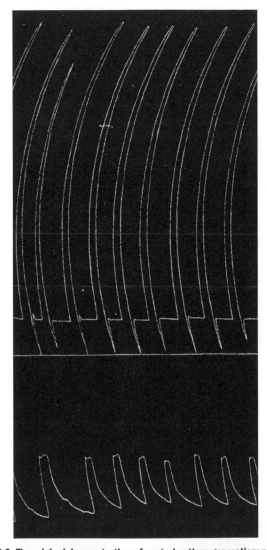

Figure 5.2. The original demonstration of central pattern generation: part of
Graham Brown's record of walking-like alternating muscle contractions in a cat
with no sensory inputs from the legs. Upward deflections in each of the two white
records indicate muscle contractions. The flexor muscle is above, the extensor
muscle below. The horizontal axis is time. The deflections are curved because
they were produced by an aluminum lever (attached to the muscle by a thread)
scraping against a rotating smoked drum.

muscles to contract and it makes their contractions alternate. Sensory inputs from the moving legs may modify this behavior—such modifications presumably would be important when a cat walks on uneven ground, for example—but such sensory inputs are not required for the basic rhythm.

Brown was the first person to demonstrate (and appropriately conclude) that a part of the central nervous system can generate rhythmic behavior by itself. When the central program is switched on (by tickling the ear in Sherrington's scratching experiment and by cutting the spinal cord in Brown's walking experiment), it generates the rhythm until it is switched off or "runs out of gas."

This finding and this conclusion were historical milestones. (Actually, as we will see, they were about fifty years ahead of their time.) It would later be shown that the basic rhythms for all types of vertebrate locomotion (walking, running, swimming, and flying) can be generated by the spinal cord.[9] (The picture is still cloudy for humans and other primates, who do not readily generate locomotion after complete spinal cord injuries. But most researchers are persuaded that the human spinal cord contains a program(s) for locomotion as well. It is just very difficult to switch the program on. There are reports of individual cases of complete or near-complete spinal cord injury in which people still make walking movements spontaneously, when their body weight is supported and they are put on a treadmill, when their nerves or spinal cord are electrically stimulated, or when they are asleep.)[10]

But Brown didn't stop there. Within the same article, he also proposed a hypothesis for *how* the spinal cord generates the walking rhythm.

Brown's proposal, now called the half-center hypothesis, went like this: There are pairs of competing "centers" (which he later called half centers instead) in the spinal cord. There could be any number of pairs, but to keep things simple, let's suppose there is just one pair for each leg: one center controls flexor muscles and the opposing center controls extensor muscles. Each center inhibits the other, which is how neurons compete. Initially (when there is no walking or scratching going on), there is kind of balance in their competition (call it a stalemate), so no movement happens.

Then, something happens to disrupt this balance (in Brown's experiment, it was cutting the spinal cord) and favor one or the other center. The favored center (let's say it's the flexor) becomes very active and triggers a (flexor) muscle contraction. At the same time, it inhibits the opposing center, so no opposing (extensor) muscle contraction is possible. But the active center gradually gets "fatigued." (What this really means is not clear, but let's just go with it.) When it gets fatigued, it becomes less effective in inhibiting its opponent.

Eventually, the opposing center overcomes the inhibition and takes over. (Brown suggested the opposing center would be helped by "central rebound" [now called postinhibitory rebound]. This was an idea that Sherrington and others had suggested to explain how, after a simple movement [like leg flexion] is triggered and its opposite [extension] is simultaneously inhibited, an opposing movement [extension, in this case] occurs spontaneously, immediately afterward.[11] This is like a ball that is thrown in one direction at a wall and then bounces back in the opposite direction.) The second center then excites its own (extensor) muscle to contract while inhibiting its opponent. Then the second (extensor) center also fatigues, and control goes back to the first center. And so on, back and forth, like a pendulum swinging, so that muscle contractions alternate rhythmically.

This was brilliant! Brown had demonstrated that the spinal cord by itself could generate a rhythmic behavior and he had even proposed a plausible way it could happen. (It would later be shown that his hypothesis is actually correct for a number of rhythmic behaviors, including a leech's heartbeat—leeches don't have heart pacemaker cells; they rely on a neuronal circuit to generate their heartbeat.[12] Ironically, though, it has been shown that a half-center mechanism in the spinal cord is *not* necessary for walking in mammals, or for several other rhythmic behaviors generated by the spinal cord.)[13]

But while Sherrington was already famous and would go on to earn a Nobel Prize and a knighthood, Brown was relatively unknown. Apparently, no one was ready to hear what Brown had to say. His papers had little impact at the time. After serving in World War I, Brown continued doing research for several years as a professor of physiology in Cardiff, Wales, but he later gave up research and turned enthusiastically to alpine mountain climbing instead, pioneering certain ascents of Mont Blanc.[14]

Fast-forward fifty years.

Another young man, an American named Donald M. Wilson, was also completing his scientific training, working with Torkel Weis-Fogh at the University of Copenhagen. Wilson and Weis-Fogh apparently knew nothing about Brown's work (or at least, Wilson didn't cite Brown's work in his articles). Wilson was working on locusts (basically, grasshoppers), not mammals. But he was interested in essentially the same question that Sherrington and Brown had been interested in: How are rhythmic behaviors generated? Wilson wanted to know how locusts fly.

Like Brown's mentor, Sherrington, Wilson's mentor, Weis-Fogh, was on record arguing that sensory neurons (such as in the base of the wings) trigger alternating reflexes to cause the rhythmic movements (of flying, in this case).[15] Wilson set out to find the key sensory nerves (which contain the axons of sensory neurons) that produce these flight reflexes. But he couldn't find them. Each time he cut a sensory nerve, the locust just kept flying.[16] Eventually, he cut all of them and the locust still flew (though a bit slower).

Wilson reluctantly came to the conclusion that sensory neurons from the wings are not necessary for flying after all. Somehow, the lower part of the locust's central nervous system (called the ventral nerve cord, instead of the spinal cord) is sufficient to generate the flying rhythm. He concluded that the locust's central nervous system contains a "central pattern generator" for flight. He published his initial results in a 1961 article, exactly fifty years after Brown's first such article, and, like Brown's, it was authored only by him.[17]

The time apparently was right in 1961. Many other examples of central pattern generators were soon described in laboratories throughout the world.[18] It turned out that central nervous systems in all sorts of animals contain central pattern generators for a variety of rhythmic behaviors, including walking, swimming, flying, scratching, breathing, chewing, calling, chirping, copulating, egg laying, and digesting. Graham Brown's pioneering discovery was rediscovered and his half-center hypothesis examined and tested (though Brown was in his eighties by then).[19]

And here's another strange coincidence: Wilson, like Brown, was a pioneering mountain climber![20] Wilson was a great counterexample to the stereotype of scientists as nerds. Before he was even fully trained as a neurobiologist, he had already been the first (along with other members of his climbing parties) to climb several challenging peaks in the American Southwest. He was perhaps even more daring and pioneering as a mountain climber than as a neurobiologist. Not quite Indiana Jones, perhaps, but Wilson was real. Perhaps Wilson was animated by the ghost of Graham Brown (though, actually, Graham Brown was still alive then).

Tragically, Wilson died at age thirty-seven during a rafting trip on the Salmon River in Idaho, which was in an extreme flood stage at the time (several other deaths occurred then, too).[21] A member of Wilson's party was ejected from his raft and carried into some brush. Wilson ran his raft aground on a sand bar and swam to reach his companion, first tying himself to his raft via a rope (perhaps a mountain climbing habit) to prevent being swept away. The river's current pulled him down and held him down via the rope, and he drowned. His ashes were scattered over the river.

Breathing Not Easy

What about breathing? It has been known since the eighteenth century that breathing in mammals *does* require the nervous system (not just the diaphragm muscle that expands the lungs).[22] In fact, it requires a very specific part of the nervous system—the medulla, which is a part of the brainstem (the elongated portion of the brain, connected to the spinal cord, that the rest of the brain sits atop like a fruit on a stem). The role of the medulla was first discovered in experiments on rabbits: the animals continued breathing when most of the brain was removed, but stopped breathing when a certain part of the medulla was removed. If this part of the medulla is damaged (or disconnected from the diaphragm, for example, by a spinal cord injury), then a person cannot breathe on their own and must be hooked up to a ventilator.

In the 1980s, Jeff Smith and Jack Feldman showed that if they cut out just a portion of a neonatal rat's central nervous system (that included the lower part of its brainstem and the upper part of its spinal cord) and placed it in a dish of oxygenated saline, it continued to produce a breathing rhythm, much like a frog or turtle heart continues to beat.[23] This showed that there is a central program that is sufficient to run the breathing factory.

The breathing rhythm in brainstem motor neurons continued after another cut that removed the spinal cord. When a slice of the medulla was cut out and placed by itself in a dish, it still generated a breathing-like rhythm for inspiration (breathing in), provided it contained a structure called the pre-Bötzinger complex (preBötC).[24] (You might guess this structure was named after a long-dead German anatomist. Actually, when Jack Feldman and colleagues were closing in on the key brain region for inspiration in 1978, discussing the as-yet unnamed structure at the banquet of a scientific meeting in Hirschhorn, Germany, Feldman picked up a bottle of white wine at the table and suggested taking the name of the wine brand, Bötzinger.[25])

So a certain part of the medulla is sufficient to generate a breathing rhythm. How does it do this?

The fact that you can take out part of the brainstem and keep it healthy in a dish for a few hours allowed researchers to work out the neurons, synaptic connections, and mechanisms involved. But the scale of this neuronal circuit is vastly greater than that of the crayfish LG escape circuit and the neurons are smaller and not individually identifiable, so this is a much harder problem to solve. Nonetheless, important progress has been made in working out the key types of neurons and synaptic connections involved.[26]

The data led to some disagreements within the field. Some said that the basic rhythm for rodent breathing is generated by a set of pacemaker neurons in the preBötC.[27] The voltage of each such pacemaker neuron would go up and down by itself, even when it receives no inputs, like a heart pacemaker cell. These pacemaker neurons would generate spikes rhythmically and so make other neurons spike rhythmically, eventually causing muscles to be activated rhythmically.

Others said that the basic rhythm for rodent breathing is generated by a neuronal circuit in which no individual neuron is a pacemaker cell. Instead, the synaptic interactions among components of the circuit generate the rhythm. One way this could happen is via synaptic inhibition, much as Graham Brown had suggested for cat walking.[28] Another way this could happen is via synaptic excitation that keeps building up until it reaches a certain point, then shuts down, starting up again in a few seconds.

So who was right? Experiments demonstrated that there *are* pacemaker neurons for breathing in the preBötC.[29] These pacemaker neurons contain a set of ion channels that together form a kind of electrical circuit that makes their voltage go up and down cyclically and continually (using a different set of ion channels from those in heart pacemaker cells). When all synaptic inputs to these neurons were blocked by drugs, their voltage continued to go up and down, up and down, at approximately the rate of normal breathing (see Figure 5.3).

So that's it, then: Breathing is generated by pacemaker cells, like the heartbeat, right?

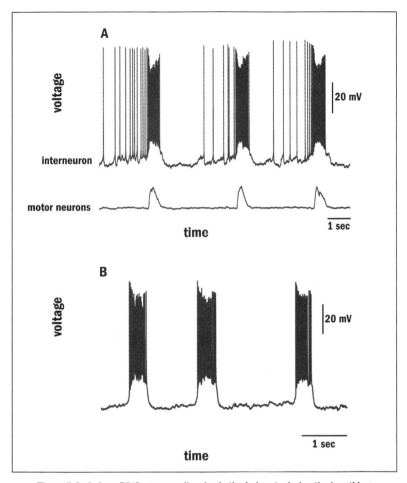

Figure 5.3. *A.* A preBötC neuron spikes in rhythmic bursts during the breathing rhythm. *B.* A preBötC pacemaker neuron continues to spike in rhythmic bursts after it has been isolated from all synaptic inputs.

Surprisingly, when these pacemaker neurons had their pacemaking ion channels shut down (using a drug), the preBötC could *still* generate a breathing rhythm (though the breathing rhythm did stop if the experiment was done somewhat differently).[30] Also, when all synaptic inhibition was blocked with drugs, the rhythm *still* continued.[31] The best guess as of 2015 is that a *combination* of factors, including pacemaker neurons, excitatory synaptic connections, and perhaps inhibitory synaptic connections, normally work together to generate the electrical rhythm that makes mammals inhale rhythmically. Also, another part of the brainstem, farther up, can generate a breathing-related rhythm on its own; this rhythm may be used for expiration (breathing out) and coordinated with the inspiratory rhythm, though mammals at rest mostly breathe using just inspiration (and exhale passively each time the diaphragm relaxes).[32]

So the bottom line is that there are multiple, redundant mechanisms that normally work together to generate the breathing rhythm.[33] These include both pacemaker neurons and synaptic interactions within a central program. But feedback from neurons that sense oxygen or carbon dioxide levels in the blood or cerebrospinal fluid modifies the program in the medulla. So there is a central program, which uses multiple mechanisms simultaneously, as well as feedback mechanisms, which provide moment-to-moment corrections. Having these multiple mechanisms probably helps keep the factory going even when there are injuries and also helps adjust the factory's output to the immediate need for oxygen.

6

The Plot (and the
Chemical Soup) Thickens

Go with Your Gut

In retrospect, perhaps it should not have been surprising that both pacemaker neurons and synaptic connections contribute to the breathing rhythm, because a similar story had already emerged from the study of a much smaller (if not simpler) central pattern generator. This neuronal circuit is found within the Rodney Dangerfield of rhythms— digestion—in lowly crustaceans.

Why would scientists study digestion in crustaceans? It's not because they just really want to know how crustaceans digest their food. It's because of Krogh's Principle (that you should choose an animal well suited to answer your particular question). This system has a number of practical advantages for working out neuronal circuitry for rhythm generation.[1]

The central nervous system of invertebrates generally consists of a set of ganglia—collections of neuronal cell bodies—and axons connecting these ganglia to each other. In crabs and lobsters, one of these ganglia, called the stomatogastric ganglion (STG), contains only about thirty neurons (depending on the species). Each of these thirty neurons can be individually identified—given a name—and the "same" neuron can be studied in many different animals of the same species. This is something that can't be done in vertebrates (except for some rare neurons like the M cell in fish). Also, each of these neurons' cell bodies is huge, so you can poke a microelectrode into not just one, but several of these thirty neurons simultaneously and monitor their electrical signals for hours. This has allowed researchers to work out the entire neuronal circuit of the STG, which would be nearly impossible to do in a mammalian brain.[2]

Now you might imagine, as most neurobiologists who study mammals did (and some still do), that a neuronal circuit that controls digestion in a lobster or crab would be so simple that it would have nothing to tell us about how mammalian nervous systems work, because mammals are so much more complex. But it turned out that the neuronal circuit controlling crustacean digestion is incredibly complex (see Figure 6.1).

First of all, the STG doesn't just generate one digestive rhythm; it generates several.[3] Crustacean digestion requires several distinct kinds of rhythmic activities at different stages (as human digestion also does). This includes peristalsis (rhythmic movement of food along the digestive tract) and chewing (these animals do their chewing in their stomachs, instead of in their mouths). About thirty STG neurons can generate four distinct rhythms that control different digestive stages (with help from neurons in other ganglia for some of these rhythms). They can also generate all kinds of intermediate and combined rhythms. This is not simple!

In fact, this neuronal circuit was so daunting that one of its leading researchers, Al Selverston, wrote an article in 1980 entitled "Are central pattern generators understandable?"[4] But all hope was not lost. Soon after, John Miller and Selverston developed a new technique to aid progress: they figured out how to delete one neuron from the circuit and study the effect.[5] They did this by injecting a neuron with a

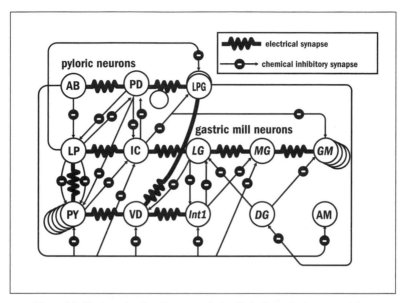

Figure 6.1. Diagram showing the neuronal circuits in the crab stomatogastric ganglion (STG) that control two different digestive rhythms, called the pyloric and the gastric mill. Neurons that control pyloric muscles are indicated in roman font, while those that control gastric mill muscles are indicated in *italics*. Two other neurons, indicated in bold font, control both pyloric and gastric mill muscles (lateral posterior gastric neuron, LPG) or a different muscle entirely (anterior median neuron, AM). See the key for indications of electrical synapses and chemical inhibitory synapses.

dye that fluoresces when a particular wavelength of light is shined on it. When they then exposed the ganglion to sustained light of this wavelength, the fluorescing neuron absorbed the energy, overheated, and died, without disturbing the rest of the circuit. With this technique, called photoinactivation, they (and others) could test the effect of each individual neuron experimentally.

This technique allowed them to test two competing hypotheses about how a particular digestive rhythm was generated: that it was determined by a pacemaker neuron (named anterior burster—AB) and that it was determined by synaptic interactions. (Sound familiar?) When they eliminated the AB cell by photoinactivation, the rhythm continued but was slowed.[6] So it appears that AB is not necessary for the rhythm, but it normally contributes to it and sets the pace. There is also a key contribution from mutual synaptic inhibition acting much like the half centers Graham Brown envisioned for cat walking.[7] In other words, there are redundant mechanisms to generate this digestive rhythm in crustaceans, just as would later be found for the breathing rhythm in mammals.

But it would be very difficult to test the role of every neuron and every synapse between neurons experimentally. And this neuronal circuit is so complicated that it is impossible to just look at it (even if you are a researcher in this field) and understand how it works. So researchers designed computer simulations (or "models") of the circuit. Then they tweaked the models to imitate all kinds of situations and see what role each neuron, each synapse, and each ion channel type plays. In effect, they performed experiments on the computer model.

In addition, Andrew Sharp and colleagues developed a cyborg-like hybrid between a crab ganglion and a computer called the dynamic clamp.[8] They poked electrodes into neurons in the living ganglion and then connected the electrodes to a computer program that simulated the behavior of a particular type(s) of ion channel. The computer program controlled injections of positive or negative current into neurons in a complex way that mimicked having more or fewer ion channels of a particular type, while the living ganglion continued to generate real-life activity. In this way, they could do precise manipulations of components of the living neuronal circuit (see Figure 6.2 for an example).

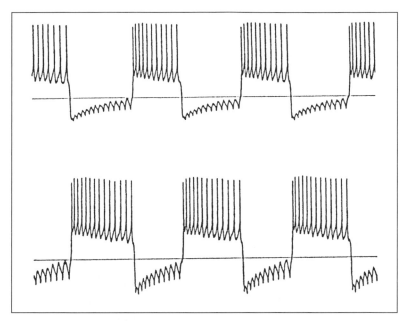

Figure 6.2. Actual intracellular recordings from two physically unconnected neurons, using the dynamic clamp to artificially add funny channels to each neuron and to connect the two via inhibitory synapses, forming a half-center oscillator.

All of this helped researchers understand the role of each neuron, synapse, and type of ion channel in generating each rhythm. But it also turned out that this neuronal circuit is not static. Anatomically, the neurons and connections shown in Figure 6.1 are always there. But under the influence of chemicals called neuromodulators, they morph into different functional circuits.[9] Neurons can change from one type to another (such as from a pacemaker-like neuron to a passive neuron), and synaptic connections can get stronger, get weaker, or effectively disappear entirely.

And it turns out that there is not just one chemical that can do this—there are *dozens*. Each has a different effect on a subset of circuit components. And the best guess is that inside the living animal the ganglion is normally drenched in a rich soup of these neuromodulators. So even with all we know about this neuronal circuit from decades of work, it is still not clear how it normally functions during the animal's life.

The complexity of the STG circuit and the flexibility induced by neuromodulators was scary for neurobiologists because it greatly increases the difficulty of figuring out how the circuit works. Is this typical of all neuronal circuits?

Understandably, researchers of mammalian brains did not want to face this nightmare. For decades, many argued that because these pesky invertebrates have such puny nervous systems, they have to make individual neurons within the nervous system do multiple jobs—be "multifunctional." But because we mammals have a wealth of neurons, we can allow each neuron to behave more simply. But as related experiments were done in mammals it became increasingly clear that mammalian neurons are every bit as complex as crustacean neurons and respond in much the same way to neuromodulators.[10]

And the crustacean ganglion has more yet to teach us. It turns out that STG circuits from individual crabs of the same species are not identical. Although the same neurons and synapses are found in each, the proportion of each type of ion channel in a particular neuron or at a particular synapse can vary dramatically.[11] Nonetheless, the STG motor neuron spiking patterns are essentially the same. So in a circuit that has redundant mechanisms to make an electrical rhythm, different crabs might actually achieve the same result in lots of different ways. In a sense, animals of the same species may be less like cars produced on an assembly line that have interchangeable components and more like custom-built cars, with each performing a car's tasks using somewhat different components.

Astrid Prinz, Dirk Bucher, and Eve Marder wondered how many ways there are to skin a cat (or to digest a crab's meal, anyway). So they changed the proportions of various types of ion channels in three stomatogastric neurons in computer model simulations—about 20 million times!—and then assessed which versions of the model produced the right output.[12] Remarkably, they found that many versions of the model worked quite well, including versions that had little in common with each other.

This could be seen as very liberating. If it applies to mammalian brains (which it almost certainly does, if it is like everything else that has been found in the STG), then we can each achieve our behavioral goals in different ways. If something goes wrong with one way (due, for example, to brain damage), we can still reach the same outcome in another way.

So we've seen that the STG is made up of complex, often multifunctional neurons that are connected in complex ways, and both these neurons and the connections among them are constantly modulated by multiple chemicals.

But that's not all. These neurons also regulate themselves to maintain a stable rhythm. For example, if you take one of these neurons out of the ganglion and put it in a dish of saline, it stops spiking rhythmically, because it's lost its synaptic inputs—it is no longer part of a circuit. But after a couple of days on its own, it spontaneously starts spiking rhythmically again![13] It essentially transforms itself into a pacemaker neuron to re-create its rhythmic spiking. If you then inject negative current into it rhythmically to mimic the inhibition it used to receive when it was in the ganglion, it becomes passive again, as it no longer needs to be a pacemaker neuron to keep the rhythm going. It apparently does this by changing its proportions of particular types of ion channels. (And in case you were thinking this is just a crustacean thing, similar things have since been found in the cerebral cortex of mammals.)[14]

But wait, it gets worse. When Jason MacLean, Ying Zhang, Bruce Johnson, and Ron Harris-Warrick physically added one type of ion channel to one of these neurons through a microelectrode, this should have changed the electrical signaling of the neuron, as the injected channels were incorporated into the neuron.[15] (The researchers actually injected the RNA that codes for this ion channel protein.) Yet the neuron's pattern of spiking was essentially unchanged, despite the fact that there really were a lot more of these ion channels functioning in the neuron. How could this be?

It turned out that the neuron compensated for the injection of these additional ion channels of one type by synthesizing more ion channels of a *different* type on its own, to balance out the net effect.[16] Pretty smart for a lowly crustacean gut neuron!

How did the neuron "know" that it needed to compensate? One might imagine that the neuron somehow monitored its own pattern of spiking (crab gut neuron self-consciousness?) and altered how many ion channels of each type it synthesized to keep on an even keel. But the problem was, there never seemed to be any change in the pattern of spiking that could be detected and compensated for.

Another experiment made the situation clearer. The experimenters injected into the neuron a *nonfunctional* version of the RNA (they included an error in the RNA code so it didn't make the right protein).[17] But the neuron still "compensated" for the injected RNA by producing more of the second type of ion channel. In this case, the neuron's compensation did *not* stabilize its pattern of spiking. So the neuron must somehow sense the amount of RNA for each type of ion channel (even though the injected RNA was a dud in this case) and try to compensate accordingly. Presumably, neurons are doing this all the time.

It also turns out that crabs exposed to warmer or colder water (which dramatically speeds or slows down, respectively, many cellular processes) compensate to maintain stable and appropriate patterns of rhythmic spiking.[18] This kind of compensation does not require synthesis of additional ion channels—it seems to be a property of the individual ion channel proteins. So individual molecules within a neuronal circuit also do a form of compensation to maintain appropriate rhythmic outputs.

For neurobiologists, there is something troubling about all these findings from the STG, because we want to fully understand how neuronal circuits generate behaviors. You could say that the neuronal control of crustacean digestion is so complex that it is almost beyond the ability of the human brain to comprehend! But if that is true of a crab's gut, what about figuring out how a human brain works? The multifunctional neurons, the effects of neuromodulators, the variability of the "same" neuronal circuit between animals, and the ability of neurons to compensate for imposed changes—in a small, well-defined circuit controlling crab digestion, which you would think would be easy—all suggest that it will be that much more difficult to fully understand how larger neuronal circuits that control more complex behaviors actually work. In a sense, we now know that neuronal circuits are moving targets, both because neuro-modulators alter the way the circuit functions and because neuronal circuits are continually modifying themselves. It's going to be a long haul to understand how mammalian neuronal circuits work, but we should probably continue to listen to the (crustacean) gut.

Are Rhythms Made by Specialists or by Jacks of All Trades?

We've seen that many neurons in the STG are multifunctional. That is, they contribute to generating more than one kind of rhythmic behavior. But how common is this across animals and across rhythmic behaviors? Is each rhythm-generating neuron a specialist who does just one thing, or is each a jack of all trades who contributes (along with many others) to a wide range of tasks?

One can imagine that there would be pros and cons to each way of organizing neuronal circuits. It would probably be easier to devise a circuit in which each neuron does just one job than to figure out a way for each neuron to do multiple jobs and yet make sure each job is done correctly. Presumably, evolution would also find such a solution more easily.

On the other hand, having each neuron do multiple jobs would be more efficient and you wouldn't need as many neurons to accomplish all required tasks. This was an argument some mammalian researchers voiced early on to dismiss the relevance (to mammals) of multifunctional STG neurons—those animals just don't have enough neurons to play with, so it works best if each neuron can do multiple jobs. But even in a large nervous system, if you have a lot of tasks to do (and you can keep learning new ones), can you really afford to dedicate sets of neurons to just one task? Each time you wanted to do a new task, you would need to somehow find more neurons. (And no, we don't use just 10 percent of our brains.)

So which is the norm in the generation of most rhythmic behaviors—specialist neurons or multifunctional neurons? To answer this question, you have to be able to study each neuron—by sticking a microelectrode into or next to it— while triggering multiple types of approximately normal rhythms. This is a challenging task that requires stability, so in practice it has been done mainly when the nervous system is not moving. That is, either the nervous system has been taken out of the animal and placed in a dish of saline or the animal has been immobilized somehow. In either case, the neuronal circuits must still produce multiple "normal" patterns of rhythmic spiking.

The story for most rhythmic behaviors examined so far is that neuronal circuits with multifunctional neurons dominate, just as neuronal democracies dominate.[19] For example, in addition to crustacean digestive rhythms, this is true for swimming and crawling in sea slugs *(Tritonia)* and leeches, ingestion (feeding) and egestion (spitting out of food) in sea hares *(Aplysia),* swimming and struggling (in which animals thrash back and forth more slowly but more forcefully than in swimming) in tadpoles and larval zebrafish, swimming and scratching in turtles (see Figure 6.3A), walking and scratching in cats, and various forms of breathing (like normal, quiet breathing versus gasping) in mammals.[20]

There are also neurons that have been shown to contribute to one (or more than one) rhythmic behavior as well as a

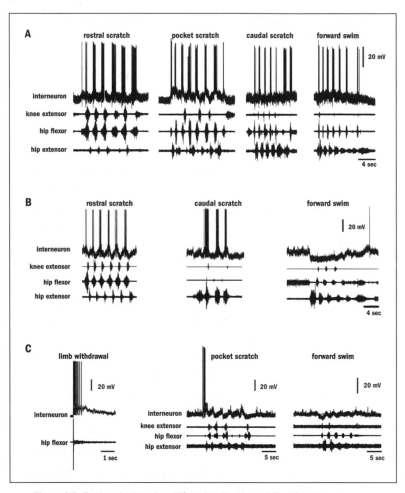

Figure 6.3. Turtle spinal cord multifunctional and specialized interneurons.
A. A multifunctional interneuron spikes rhythmically during all three forms of
scratching (which the turtle uses to reach three regions of the body surface)
and swimming. *B.* A scratch-specialized interneuron spikes rhythmically during
scratching and is inhibited (its voltage goes down and it does not spike) during
swimming. *C.* A limb withdrawal–specialized interneuron spikes during limb
withdrawal and is inhibited rhythmically during the scratching and swimming
rhythms.

nonrhythmic behavior, such as swimming and shortening in leeches; swimming and escape in larval zebrafish; and swimming, scratching, and limb withdrawal in turtles.[21]

But this is not the whole story. There are also behaviorally specialized neurons and even behaviorally specialized circuits.[22] For example, separate circuits of interneurons generate walking and flying in locusts, as well as chirping and flying in crickets, even though these separate sets of interneurons produce these behaviors through connections to the same set of motor neurons and muscles.[23]

An emerging theme from several neuronal circuits is that a combination of multifunctional and specialized neurons contributes to these behaviors.[24] For example, most neurons active during swimming are also active during crawling in leeches, but some are only active during crawling.[25] One might say they are specialized for crawling. Similarly, most interneurons that generate ingestion in the sea hare *Aplysia* also contribute to egestion, but some are specialized for one or the other.[26] In vertebrates, most spinal cord interneurons that contribute to struggling in tadpoles and larval zebrafish also contribute to swimming, but some are struggling specialized.[27] Some zebrafish neurons are escape specialized.[28] Most spinal cord interneurons that spike rhythmically during scratching in turtles also spike rhythmically during swimming (see Figure 6.3A) and spike during limb withdrawal.[29] But others are scratching specialized and are inhibited during swimming (see Figure 6.3B).[30] Still others are limb-withdrawal specialized and are inhibited during both swimming and scratching (see Figure 6.3C).[31]

The bottom line is that circuits that generate rhythmic (and some other) behaviors often use a combination of multifunctional and specialized neurons. Apparently, neuronal circuits like to use jacks of all trades, but some jobs also require specialists. The different roles played by multifunctional versus specialized neurons in these circuits are still being worked out.

Picking Teams

If a neuron's anatomical connections would allow it to contribute to multiple behaviors, what makes it spike during one rhythmic behavior versus another? In other words, how do neurons decide which behavioral team to play for? And how does a circuit full of multifunctional neurons decide which game to play?

One answer, as we've seen, is that it depends on the neuromodulatory environment—the chemical soup that bathes the neuron. In some cases, you can turn on a particular rhythm by adding a certain chemical or chemicals to a part of a nervous system sitting in a dish of saline. This chemical determines which neurons in the circuit spike and when. For example, adding dopamine to a leech nervous system triggers the crawling rhythm.[32] Adding serotonin to a male frog brainstem triggers the calling rhythm.[33] Adding serotonin and a drug called NMDA (which attaches to one type of glutamate receptors) to a rodent spinal cord triggers a walking-like rhythm.[34]

But how does this relate to what normally happens in the living animal? In some cases, spikes in sensory neurons (or in interneurons that are excited by sensory neurons) can trigger the secretion of a neuromodulator that in turn triggers the rhythm. This happens, for example, in the sea slug (*Tritonia*). In this case, the chemical secreted (serotonin) modifies chemical synapses made onto the very neuron that secreted the chemical (among others).[35] This is a neuronal version of pulling yourself up by your own bootstraps (or perhaps picking yourself for your own team).

In other cases, though, the neuronal circuit will be bathed in some particular neuromodulatory soup at any given time, but this may merely create a context or "mood" that modifies a rhythm if the rhythm starts. Additional spikes in certain neurons may still be required to trigger the rhythm. In such cases, what triggers the rhythm? And what determines which rhythm is generated in a multifunctional circuit?

Let's look at two examples. First, leeches.

Leeches mostly use multifunctional neurons to generate two forms of locomotion: swimming—a wavelike movement through the water—and crawling—an inchworm-like movement along the substrate, such as the bottom of an aquarium (see Figure 6.4).[36] When the water is deep enough, leeches prefer to swim, which is faster. But if the water is too shallow, leeches crawl instead. At an intermediate water level, they will swim some of the time and crawl some of the time, unpredictably. The leech's nervous system will also produce either a swim-like or a crawl-like pattern of motor neuron spikes if it's pinned down in a dish of saline and a skin sensory nerve is shocked.

Kevin Briggman, Henry Abarbanel, and Bill Kristan devised a way to monitor the spiking of lots of leech neurons while the nervous system decides whether to swim or to crawl.[37] It would be prohibitively difficult to poke microelectrodes into one hundred neurons simultaneously, but they were able to use a pair of dyes that fluoresce when each neuron's voltage changes (one dye fluoresces when the voltage goes up, the other when the voltage goes down, and they measured the ratio of these two). They bathed a whole ganglion in these two dyes and then monitored the electrical

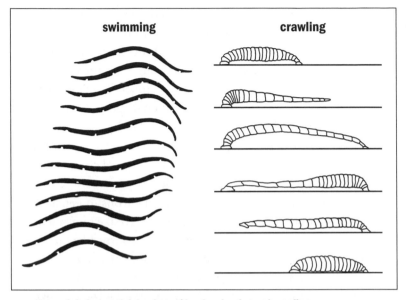

Figure 6.4. Sequential drawings of leech swimming and crawling.

signals (via the fluorescence) of more than one hundred neurons simultaneously.

At the time Briggman and colleagues did these experiments, they already knew a lot about which neurons in the ganglion (each of which has a name—or actually a number—and can be found in each member of the species) are multifunctional and which only spike during crawling, for example. At first, they looked to see which individual neurons could predict earliest whether the leech would swim or crawl (it takes a couple of seconds for the swim or crawl to get going, and they focused on this decision-making period).

Then they also looked at which *combinations* of neurons could predict the behavioral choice earliest, using sophisticated statistical techniques (principal components analysis and linear discriminant analysis) to assess how voltage changes covary among sets of neurons before a swim or crawl starts. They found that combinations of neurons could predict the behavioral choice earlier than individual neurons—and here's the strange part: the combinations that predicted the choice earliest did *not* include any of the individual neurons that predicted the choice earliest.

The events that occur earliest are most likely to be making the decision. This means that the leech's decision whether to swim or to crawl might be made through correlations in the voltage changes of a set of neurons (for example, swim whenever neurons A and B spike and neurons C and D don't spike), rather than the spiking of any one neuron.

Briggman and colleagues then focused on individual neurons within the group that predicted the behavioral choice earliest. One of these neurons was an old friend—cell 208—that had been identified long before as a neuron that spikes rhythmically during swimming. So it seemed likely that they could bias the leech's decision toward swimming by injecting positive current into cell 208 to make it spike more during the time that the leech was deciding. What they found was the *opposite:* injecting current to make cell 208 spike more increased the probability of *crawling* (and *decreased* the probability of swimming).

So the lesson here is that a group makes the decision, not a single neuron, and you can't necessarily tell what contribution one neuron makes to this group decision just by looking at the spiking of that one neuron on its own—you have to look at it in the context of what other group members do.

This story causes concern about whether we can figure out how a network composed of a combination of multifunctional and specialized neurons decides which rhythmic

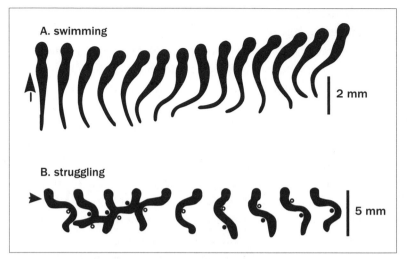

Figure 6.5. *A*. Tadpole swimming. *B*. Tadpole struggling. These are tracings from sequential frames of a high-speed camera (200 frames/sec).

behavior to generate by studying one neuron at a time. But let's not give up hope. Let's instead turn to another example: swimming and struggling in tadpoles.

Wenchang Li, Alan Roberts, Steve Soffe, and colleagues studied how very young tadpoles of African clawed frogs *(Xenopus)* generate movements. At this early stage, there are only about ten kinds of neurons in their spinal cord, which makes it much easier to work out the neuronal circuits.[38]

If you touch a young tadpole's skin briefly, it rapidly swims forward, using a wavelike motion—about twenty waves per second—with the wave moving from the head to the tail (see Figure 6.5A). If you instead grasp the tadpole, it makes a stronger, wavelike struggling movement—about five waves per second—with the wave moving from the tail to the head (see Figure 6.5B). So these are clearly different behaviors. For both behaviors, muscle contractions alternate between the left and right sides in each segment of the body.

It has turned out that tadpoles are able to generate swimming and struggling appropriately by using a combination of multifunctional and specialized spinal cord neurons.[39] The basic circuits are outlined in Figure 6.6.

One type of skin sensory neurons, Rohon-Beard cells (RBs), triggers both swimming and struggling—it is just a matter of how many consecutive spikes these neurons generate, which depends on how long they are stimulated. Brief RB activity causes a spike in descending interneurons (dINs) that triggers swimming. dINs make both electrical and excitatory chemical synapses onto other dINs. So when a few get excited, they get their fellow dINs excited, like a neuronal pep rally. This makes many of them generate a spike almost simultaneously. The dINs excite motor neurons (and other neurons) on the same side of the body, which in turn make the muscles contract.

In the presence of the chemical glutamate (which the dINs secrete at their synapses—so this is another example of pulling yourself up by your own bootstraps), dINs act like pacemaker neurons—their voltage goes up and down cyclically, in time with the swimming rhythm.[40] So pacemaker neurons contribute to generating the swimming rhythm.

In addition to the pacemaker neurons, there is a set of synaptic connections that also contributes to generating the rhythm. It involves inhibition between the left and right sides of the spinal cord, in something like the way Graham Brown proposed for flexor and extensor centers (his half-center hypothesis). The dINs excite "commissural" inhibitory interneurons (cINs) that send their axons to the opposite side of the spinal cord (for example, from the left side to the

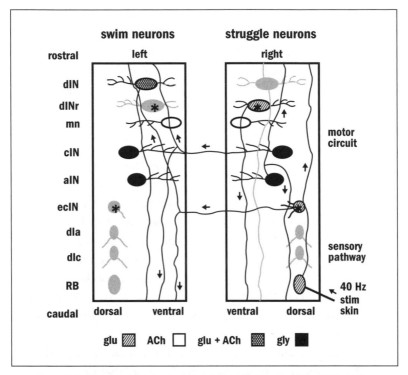

Figure 6.6. Diagram of the neuronal circuits for tadpole swimming and struggling. Types of interneurons active during each behavior are shown in black, hatching, or cross-hatching, with swimming neurons shown on the left and struggling neurons shown on the right (though these kinds of interneurons on both sides are active during each behavior); neurons shown in gray are not active during the rhythmic portion of the behavior. Rohon-Beard cells are the sensory neurons that trigger each behavior. cINs provide left–right inhibition during each behavior. dINrs and ecINs are active during struggling but not swimming; dINs are active mainly during swimming. The short branching lines from each neuron's cell body are dendrites; the longer lines are axons; arrows indicate the direction of signal flow during struggling. Black, white, hatching, and cross-hatching indicate which neurotransmitter(s) each type of neuron uses (see key at bottom). (dIN = descending interneuron; dINr = repetitively firing descending interneuron; mn = motor neuron; cIN = commissural interneuron; aIN = ascending interneuron; ecIN = excitatory commissural interneuron; dla = dorsolateral ascending interneuron; dlc = dorsolateral commissural interneuron; RB = Rohon-Beard cells.)

right), where they inhibit dINs (and other neurons). Following this brief inhibition, each (right-side) dIN spikes again, due to ion channels that mediate what Sherrington called central rebound (and is now called postinhibitory rebound). Each (right-side) dIN spike excites (right-side) cINs that in turn inhibit (left-side) dINs. And so these events keep repeating rhythmically, like a pendulum moving back and forth.

So tadpole swimming, like crab digesting and probably mammalian breathing, generates a rhythm through redundant mechanisms that include both pacemaker neurons and synaptic interactions. The left–right synaptic inhibition (half center) is probably the more important mechanism, because when all the left-side neurons were suddenly inhibited via light-activated ion channels, swimming stopped.[41]

So what causes the circuit to produce struggling instead of swimming?

The RBs also excite a type of struggling-specialized interneuron called excitatory commissural interneurons (ecINs) (see Figure 6.6). Each RB input excites an ecIN only a little, not enough to trigger a spike. But if the ecIN receives a series of inputs at a high rate (which happens when the animal is grasped and the RBs fire repeatedly, but not when it is briefly touched and the RBs fire just once), it adds them together until its voltage goes high enough to trigger several spikes. The ecINs in turn excite several types of neurons on the other side (left–right) of the spinal cord, including the inhibitory cINs, which creates another left–right half-center competition, but at a slower rate than during swimming. The ecINs also excite another type of struggling-specialized interneurons, called repetitive-firing descending interneurons (dINrs). The dINrs, unlike the dINs that are active in swimming, generate several spikes per cycle and so contribute to making each struggling cycle last longer than each swimming cycle. This creates a slower rhythm with stronger muscle contractions for struggling.

In other words, the two kinds of behaviorally specialized neurons tip the balance to make the multifunctional circuit operate in struggling mode rather than swimming mode. So you need both specialized and multifunctional neurons working together to generate each type of rhythmic behavior at the right time.

So the governance of the factories of life—at least, the governance of the rhythmic skeletal muscle contractions that underlie breathing, walking, and other crucial movements—is immensely complicated. Neither a central program nor a feedback mechanism works in isolation. Instead, both contribute to running the factory, working hand in hand. Within the central programs, both autonomous pacemaker neurons and interacting neuronal circuits contribute to running the factory. Neuromodulators can make each component operate differently, which can make the circuit output (and the behavior) very different. Neurons can also adjust their own operation to keep their outputs stable. How all these effects work together appropriately is a major question for future neurobiology research.

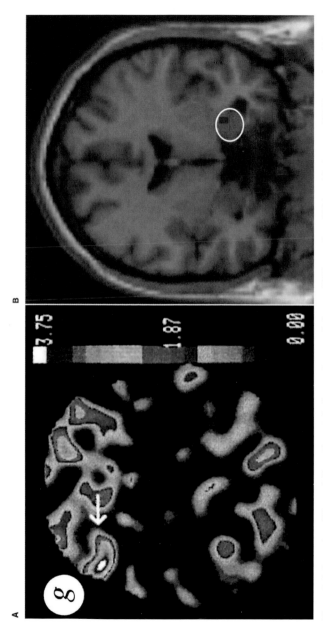

Figure 2.2. Examples of the evolution of human brain scan images. *A* shows a PET image from 1988 and *B* shows an fMRI image from 2002.

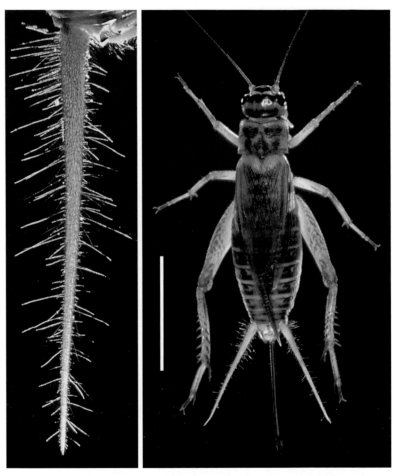

Figure 4.2. A paired structure on a cricket's abdomen, the cercus contains hairs that are each sensitive to air currents from a different direction.

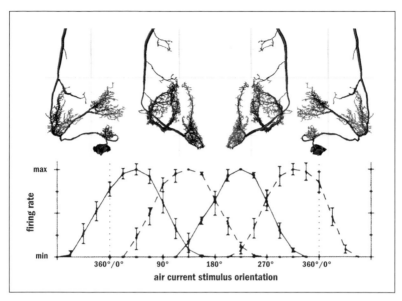

Figure 4.3. Rate of spikes in each of the four broadly tuned cricket abdominal sensory interneurons in response to air currents from different directions (with a tracing of each neuron above its tuning curve).

Figure 7.1. A barn owl.

Figure 7.4. A star-nosed mole (top) and a close-up of its star from scanning electron microscopy (bottom).

Figure 7.5. A mustached bat *(Pteronotus parnellii).*

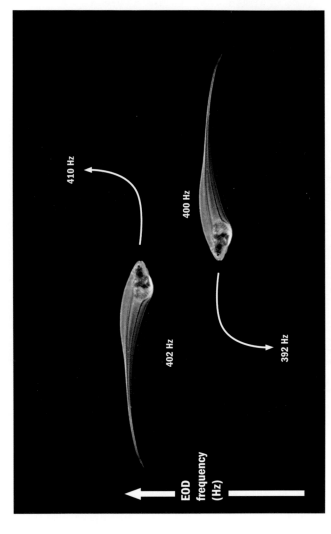

Figure 7.9. Illustration of the jamming avoidance response (JAR) of weakly electric fish, in which each fish changes the frequency of its electric organ discharge (EOD) to make it more different from its close neighbors.

Figure 9.1. A breeding pair of zebra finches *(Taeniopygia guttata).* The male is on the left. Bird Kingdom, Niagara Falls, Ontario, Canada.

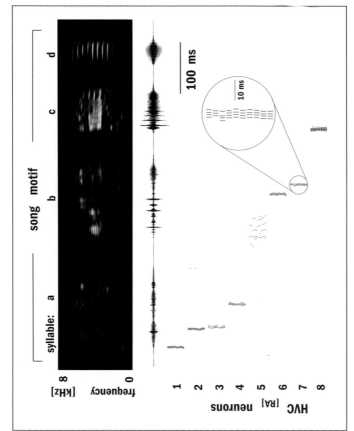

Figure 9.3. RA-projecting HVC neurons (bottom, labeled 2–8) each generate spikes (colored vertical lines) briefly and consistently (multiple trials are stacked) at one particular moment within the bird's song motif (top).

7

Government Surveillance

Governing a country well requires a vast amount of information. Governments need to keep track of what's going on both inside and outside the country.

Similarly, nervous systems have to gather vast amounts of information about what's going on inside and outside the body, to map out opportunities and dangers.

For example, nervous systems gather information to assess the body's energy needs and the energetic cost of each physiological process. This allows them to allocate resources such as glucose and oxygen (via blood flow) to various organs as needed. If more total blood flow is needed, the force and rate of heart contractions are increased. If more total oxygen is needed, breathing rate and depth are increased. In addition, if more total energy is needed, hunger is triggered.

Nervous systems monitor both internal dangers and external threats via many kinds of sensors—their sensory neurons. For example, pain sensory neurons deep inside the body signal potential physiological failures in organs. Pain sensory neurons across the skin surface signal potential tissue damage caused by external sources, such as predators. Food provides an excellent opportunity to acquire resources but can also be a threat. Sweet and salty taste sensory neurons on the tongue signal food that is high in energy content and required electrolytes. Bitter taste sensory neurons on the tongue and even in the lungs signal potential threats in the food and the air.

Many types of external opportunities and threats can be monitored from a safe distance, before a body comes into contact with them. The senses of smell, hearing, and sight usually play the lead roles in acquiring such information. Food (or prey) and predators often have distinctive odors or make distinctive sounds. If the food or predator is near enough and there are no obstacles, sight can tell the nervous system what and where it is. We humans typically rely largely on sight to obtain a precise and panoramic view of the world around us. Yet blind people often achieve a kind of "vision" of the world through other senses, such as hearing and touch. Some animals normally "see" their world through another sense, instead of or in addition to sight. These can be senses that we know from our own experience, such as hearing and touch, which have evolved to a much higher degree of precision in certain species. Two such examples are hearing in barn owls and touch sensitivity in star-nosed moles. They can also be senses that are unfamiliar to humans, such as echolocation in bats and electrolocation in weakly electric fish.

One way that nervous systems gather information is by using transparent or overt sensors. For example, when an animal smells, hears, or sees a food resource or a predator, other organisms in the vicinity generally have access to the same odors, sounds, and sights; in this sense, it is public information. When a barn owl hears the rustle of a mouse or a star-nosed mole feels a worm, they are acquiring information that is essentially publicly available, but they are doing so with a rare degree of precision.

Another way that nervous systems sometimes gather information is covertly, essentially obtaining information that is not publicly available. They may be able to use this additional information to identify opportunities and threats that are not generally known to other organisms in the vicinity.

Some species have evolved specialized surveillance devices, which generate signals that most other animals are unaware of. Many bats, dolphins, and whales image their surroundings by making sounds and analyzing the resulting echoes; this type of surveillance uses the same principles that humans use in radar and sonar imaging. It is also a more sophisticated version of what a blind person does by tapping with a cane. The precision of echolocating bats' auditory surveillance rivals what other animals, like humans, can achieve using vision. Weakly electric fish, like bats, generate a signal that is altered by objects in their environment. But these fish generate an electric field that surrounds them and then monitor where, when, and how this field is disrupted by nearby organisms or objects.

Animals that specialize in one of these types of surveillance provide fascinating opportunities for us to understand how the immediate environment can be precisely and often panoramically sensed through nonvisual means. They also allow us to search for common mechanisms that all kinds of nervous systems use to spy on the world around them.

The Wisdom of Owls

Let's begin with some animals that use senses we are familiar with, like hearing and touch, but that use them with unexpected precision, similar to how we use vision. For example, some kinds of owls can locate objects using their hearing with almost as much precision as when using sight. This unusual ability allows them to hunt at night, even when there is no moon. A barn owl (Figure 7.1) can catch a mouse in complete darkness by listening to the rustling sounds it

Figure 7.1. A barn owl. See color insert.

makes as it runs, as Roger Payne first demonstrated beginning in the late 1950s. In one experiment, Payne tied a dry leaf to a mouse's tail and allowed it to run on sand in the dark.[1] The owl struck at the leaf, rather than the mouse, demonstrating that it was the sound that told the owl the location. Masakazu (Mark) Konishi later used an infrared camera to produce the sequential images shown in Figure 7.2 of a barn owl catching a mouse in complete darkness.

In the late 1970s, Eric Knudsen and Mark Konishi poked microelectrodes into the barn owl's brain and found that there are neurons that spike in response to sounds only if they come from a particular location in space.[2] Different neurons respond to sounds from different locations. In addition, these neurons are anatomically arranged in a highly

Figure 7.2. Photo sequence of a barn owl catching a mouse in the dark, filmed with an infrared camera, from the work of Masakazu (Mark) Konishi.

organized fashion in a part of the owl's midbrain called the inferior colliculus. Moving the microelectrode along one axis systematically changes the horizontal location (azimuth) of the sounds neurons respond to; moving the microelectrode along another axis, perpendicular to the first, changes the elevation of the sounds neurons respond to. In other words, the owl's inferior colliculus contains a "map" of sound locations. The location of spiking neurons within the map signifies the location of a sound source relative to the owl. Neighboring locations in the brain respond to sounds from neighboring locations in the world.

Similar brain maps of space had been known for some time for the senses of sight and touch in many species. In general, brains form multiple maps of the space around and on the surface of the animal, using one or multiple kinds of sensory inputs. These maps appear to be critical for animals to keep track of the locations of important objects and events. So the finding of a sensory map in the owl's brain was not by itself unusual. But it was remarkable that a map of space could be generated in the owl's brain from *sounds.* In vision, the location of a nearby object (such as another animal) is provided by the location on the retina that is stimulated by light reflected from the object. Similarly, the location on the body that is touched is directly determined by sensory neurons in the skin that send signals to a map in the brain. But the location of a sound source is *not* provided by the location in the cochlea (the organ inside the ear in which sensory neurons respond to sounds). The location in the cochlea instead signals the frequency or pitch of the sound neurons respond to. So how does an owl's brain determine where a sound came from?

It turns out that the owl's brain uses two separate systems to analyze two different kinds of information that can be extracted from the sound. One system gathers information on the precise time at which the sound activates the left ear versus the right ear. The relative timing of the sound at the two ears gives the brain the data it needs to determine the sound's azimuth. A sound coming from the owl's left side hits the left ear first; a sound from the right side hits the right ear first; a sound from directly in front hits both ears at the same moment.

There are neurons in a part of the barn owl's brain called the nucleus laminaris (which is earlier in the pathway than the inferior colliculus) that receive signals (through their synapses) that originate in both ears (see Figure 7.3).[3] These neurons don't respond much to signals from one ear alone. They spike more when they receive signals from the two ears at exactly the same moment, so they are called coincidence detectors. They are also called combination sensitive, because they respond selectively to the combination of two particular kinds of stimuli. (The pinnacle of combination-sensitive neurons would be a "grandmother cell"—see Chapter 3.)

A sound coming from the owl's left side reaches its left ear first, so you might think that left-ear inputs would reach all nucleus laminaris neurons earlier than right-ear inputs from the same sound; in that case, these neurons would not spike much, because they require two coincident inputs. But the lengths of axonal paths from the two ears to nucleus laminaris neurons vary systematically, forming what are called delay lines (see Figure 7.3). At one end of the nucleus, the path is longer from the left ear; at the other end of the nucleus, the path is longer from the right ear. For any particular sound source location on the owl's left side, there are neurons in one location within the nucleus laminaris for which the extra time that it takes for sound to reach the right

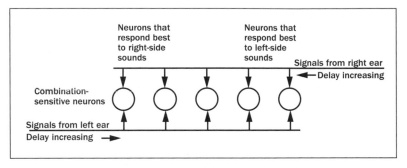

Figure 7.3. Schematic illustration of how the barn owl's brain calculates sound source azimuth. Signals from the left ear and the right ear converge onto individual neurons in the owl's brain, with the pathway being longer from the left ear for some neurons (right side of diagram) and longer from the right ear for other neurons (left side of diagram). Neurons receiving inputs that take a longer path from the left ear, for example, receive coincident left–right inputs only when the sound is produced on the owl's left side.

ear as compared to the left ear (which is just a fraction of a millisecond) is exactly balanced by the extra time it takes spikes to be conducted along the longer axonal path from the left ear (as compared to the right ear) to the nucleus. As a result, neurons in this location receive left-ear and right-ear signals simultaneously only for sounds from this particular left-side azimuth. Other neurons in this nucleus receive inputs via different left–right delay lines and so respond to sounds that come from different azimuths. These neurons are systematically arranged along one axis of the nucleus to form a map of sound azimuth. This delay-line algorithm was actually proposed for determining sound azimuth by Lloyd Jeffress in 1948, but was not experimentally demonstrated until this research on owls in the 1980s.[4]

A second brain system gathers information on the elevation of sounds by comparing the loudness or intensity of each sound at the two ears.[5] How could that possibly work?

Why would elevation affect the intensity of sound at one ear versus the other? It turns out that the barn owl's ears and the stiff feathers (facial ruff) surrounding them are asymmetric. The ruff reflects more sound from above into the right ear and more sound from below into the left ear.[6] So sounds from above are louder in the right ear and sounds from below are louder in the left ear. There are neurons in another structure in the owl's brain that respond selectively to sounds that are louder in one ear than the other, with a particular difference in sound intensity between them.[7] These neurons signal the elevation of each sound.

In the inferior colliculus, which is a brain structure farther along in the pathway, the signals indicating sound azimuth and sound elevation are combined. Individual combination-sensitive neurons in the inferior colliculus spike most when sounds come from a particular azimuth *and* a particular elevation. In other words, these neurons signal a particular location in two-dimensional space (or a particular angle from the owl). These neurons are anatomically arranged within the inferior colliculus to form a map of the spatial locations of sounds, or sound space.

Neurons in this initial map of sound space in the inferior colliculus each prefer sound of a particular frequency (or pitch) or a narrow range of frequencies. This is a potential problem, because the owl's azimuth calculation doesn't just depend on when the very beginning of a sound reaches the left ear versus the right ear. Instead, it depends on the ongoing time difference between the two ears.[8] How can the owl tell that there's a difference in when the sound reaches the two ears while the sound continues in both ears? A sound of a particular pitch is actually a wave of increasing and decreasing air pressure that repeats cyclically at a rate given by its frequency. For example, a 1,000-Hz sound is an air-pressure wave that goes up and down 1,000 times per second. The neurons in the nucleus laminaris and in the first inferior colliculus map actually signal the difference in arrival times of each sound wave peak between the left and right ears.

The problem is that this signal is ambiguous: the owl can't be sure what the sound source azimuth is unless it knows which peak at the left ear corresponds to which peak at the right ear. If the owl compares the peak of one wave at the left ear and a different wave at the right ear, it will get the wrong answer. In fact, when a sound containing a narrow range of frequencies is played to an owl in the dark, it often turns its head in the wrong direction.[9] But when a sound that contains a wide range of frequencies is played to the same owl, it always turns its head in the correct direction. How does the owl do this?

There is a second brain map of sound space (also in the inferior colliculus) in which each neuron receives inputs from multiple neurons in the first map that signal the same sound source location but use different ranges of sound frequency.[10] When the inputs from different frequency channels are combined, the ambiguity in azimuth disappears. Essentially, this part of the brain combines information from multiple sources. Information from a single source is unreliable, but when multiple sources agree, the conclusion is much more likely to be accurate.

In the tectum (called the superior colliculus in mammals), also in the midbrain but one stage later in the pathway, this second map of sound space from the inferior colliculus is merged with a similar map of visual space.[11] Each neuron in the tectum responds selectively to sights and sounds from one location in space, so the tectum's map is a multisensory map. By combining visual and auditory data, the map in the tectum gives the owl a still more reliable indication of where something is in the world.

For an owl to use both sights and sounds to capture a mouse (in the moonlight, for example), the visual and sound space maps in the tectum must be precisely aligned with each other. How do they become precisely aligned? Eric Knudsen and his colleagues explored this question by out-fitting owls with custom-made goggles containing prisms.[12] The prisms shifted the azimuth of all visual inputs without affecting sounds. Juvenile owls that grew up wearing these goggles turned their heads precisely toward the location of a seen or heard object as long as they were wearing the goggles, but made errors if the goggles were removed. Clearly, the owls had somehow adjusted to the presence of the prism goggles, but how?

When the experimenters poked microelectrodes into the tectum of these owls, they found that the neurons spiked in response to visual stimuli and sounds from the same location only when the goggles were on. If the goggles were taken off, the visual and auditory maps in the tectum were mis-aligned. It turned out that when the young owl wears the goggles, the visual map in the tectum essentially "instructs" the auditory map in the inferior colliculus to shift anatomi-cally, over a period of months, so that the two maps in the tectum are realigned. So the owl uses multiple kinds of sen-sory information to monitor locations of objects of interest, but the visual information helps set up the organization of the auditory information.

If adult owls wear the same kind of prism goggles as the juveniles for the same amount of time, however, there is little shift in alignment of the sound space map.[13] Old owls do not seem to learn this new trick well. But they can learn it better with help. If adult owls wear goggles containing weak prisms that only shift visual inputs a little (training goggles?), they actually shift their auditory space maps more. If they wear a series of goggles with incrementally increasing prism strengths, they shift still more.[14] So stepwise training makes learning much easier for adult owls. Once they have learned this large shift, if the goggles are taken off for months (and the brain maps return to their original alignments) and then the high-strength goggles are put back on, the owls quickly make the large shift again. So once the adult owls learn to "ride this bicycle," they preserve some long-term brain record of the knowledge. Also, if adult owls wearing prism goggles for the first time are allowed to hunt live mice, then they make a large shift in sound map alignment directly.[15] So something about being able to hunt live prey facilitates this kind of learning.

One lesson from the barn owl example is that not only vision can be used to create a precise and panoramic "image" of objects in the world. Hearing can also accomplish this, with the right sorts of ears and brain pathways. With both vision and hearing, animals create maps of space in their brains that allow them to move toward prey and away from predators. The combined visual and sound space map in the tectum was discovered in owls, but appears to occur widely (though not as precisely) in many vertebrates. So a second lesson may be that by studying a species that does a trick (in this case, sound localization) particularly well, researchers may make faster progress in figuring out the secret of the trick.

The Star of Touch

What about the sense of touch, the somatosensory system? Not surprisingly, locations touched on the body are indicated by spiking in sensory neurons in the skin. In mammals generally, these somatosensory signals are anatomically organized into brain maps of the body, both in the cerebral cortex and in the mammalian equivalent of the tectum, the superior colliculus. In general, however, these somatosensory maps of the body surface are not nearly as precise as maps of visual space. There are some species, however, that specialize in somatosensory surveillance and have taken it to a whole new level. The star of somatosensory intelligence gathering is the star-nosed mole, and its story has been worked out largely by Kenneth Catania at Vanderbilt University.[16]

The star-nosed mole relies almost entirely on its somatosensory system to locate and identify prey, which are mostly tiny worms and insects living in marshy soil. You might think that vision would be a better way to go, but these animals can find, identify, and ingest suitable prey in about a fifth of a second, faster than any other known mammal using any sense. This allows them to maximize their rate of acquiring resources. How do they do this?

These are very strange-looking animals (see Figure 7.4). They have what looks like a tiny starfish affixed to their nose, except that it consists of twenty-two mobile, fingerlike appendages (eleven on each side). Each appendage is packed with sensory neurons that respond to very light touch. Spikes from these somatosensory neurons are sent to multiple locations in the brain. There are somatosensory maps of the star in both the cerebral cortex and the superior colliculus. Each neuron in the cerebral cortical map spikes in response to light touch of one precise location on one of the star's

Figure 7.4. A star-nosed mole (top) and a close-up of its star from scanning electron microscopy (bottom). See color insert.

appendages.[17] This brain map looks just like the star itself: each of the appendages has its own area within the map and the appendage representations are arranged within the brain like the corresponding appendages on the animal's face.

These somatosensory maps essentially take the place of visual maps, because the star-nosed mole has very poor vision. In fact, the parallels to vision go beyond the high sensitivity and spatial precision of neurons in these maps. In visual mammals, including humans, a region in the center of each retina (the fovea) is specialized for precise identification of objects. Light reflected from an object hits this region of the retina when we look directly at the object. The photoreceptors (our visual sensors) in the fovea are especially densely packed. Within visual brain maps, neurons that respond to foveal stimulation are activated by a much smaller range of retinal locations than neurons that respond to other parts of the retina, which makes their signals more precise spatially. In addition, the part of the brain map that represents the fovea is proportionally much larger than parts of the map representing other retinal locations, which presumably allows the brain to perform finer-scale analysis of images that we look at directly.

Similarly, in star-nosed moles, there is a part of the star, near the center, that acts like a "somatosensory fovea."[18] Signals from this region are sent to a greatly expanded part of its cerebral cortical map. This presumably allows the mole to analyze somatosensory information with extra precision when signals come from this part of the star, which is activated when the animal "directly feels" an object.[19] The general rule, both for vision in most mammals and for touch in star-nosed moles, seems to be: use one array of sensors to quickly locate something of interest and use another, more specialized array of sensors to identify exactly what the object is.

So both hearing and touch can essentially take on the role of vision in generating a precise and panoramic picture of the nearby environment in certain species.

Precision Sonar

Many species of insect-eating bats, similar to barn owls, are champions of auditory precision. Bats are not blind (despite the saying), but they don't see very well and, like owls, they typically hunt at night. But bats don't rely on the sounds made by their prey. Instead, they make their own sounds, which bounce off the prey and return to the bat. This is essentially sonar, which ships and submarines use to map objects in the water. It is also analogous to radar, which follows the same principle, but uses radio-frequency electromagnetic waves instead of sound waves. Among animals, this is a relatively specialized type of covert surveillance and was not discovered until the twentieth century.

It had been known for a few centuries that insect-eating bats rely on sound to navigate their surroundings. The evidence was that blinded bats could avoid obstacles while flying, but bats with their ears plugged could not.[20] But how could bats use sounds to find their way? The bats didn't seem to be making any sounds. But H. Hartridge hypothesized in 1920 that they might make very high-frequency sounds and use the resulting echoes to form "sound pictures."[21] The problem was, there was no way to test this hypothesis at that time.

In 1937, however, a biology undergraduate at Harvard University, Donald R. Griffin, heard that a Harvard physicist, George Washington Pierce, had just invented a device that could detect much higher frequency sounds than humans can hear (above 20,000 Hz—or 20 kHz—which is called ultrasound). Griffin suggested that they try out the device on bats. They found that insect-eating bats do emit sounds much higher than we can hear (including sounds over 50 kHz).[22] The bats emit these sounds from their larynx, much as humans do when we speak or sing. Griffin and colleagues later found that such bats emit ultrasonic sounds while they fly around obstacles and speed up the sound pulse rate in the final stage of chasing down an insect.[23] Griffin named what these bats do "echolocation."

So it was apparent that these bats find their way around and catch insects by emitting high-frequency sound pulses and listening to the echoes, but how does this work? There are several kinds of information that can be extracted from the combination of a sound pulse and its echo. The bat's brain uses separate pathways to analyze these data, much as the barn owl does for sound azimuth and elevation.[24] The bat's brain can extract from echoes the distance, relative velocity, flutter (something that moths do a lot), and even texture of objects that reflect a sound.

The delay between the emission of a sound pulse from the bat's mouth or nose and the reception of its echo at the bat's ear tells the bat how far away the object is. Sound travels at approximately a constant rate in air, so half of the delay (because the sound has to travel to the object and then back) multiplied by the speed of sound yields the distance of the object. The bat's brain does the math. The calculation is implemented by combination-sensitive neurons, first in the inferior colliculus and later in the cerebral cortex. These neurons respond selectively to a pulse plus an echo with a particular delay between them.

Figure 7.5. A mustached bat *(Pteronotus parnellii)*. See color insert.

For this to work, the pulse and the echo have to be somehow different, so the bat can tell them apart. The discrimination between pulse and echo is accomplished in different ways in different species. In some, it is based on sound intensity (loudness), which is greater for pulses than for echoes (because only a fraction of the outgoing pulse energy is reflected back to the bat from each object). In others, such as the mustached bat (see Figure 7.5), it is based on sound frequency. The mustached bat has been used for in-depth

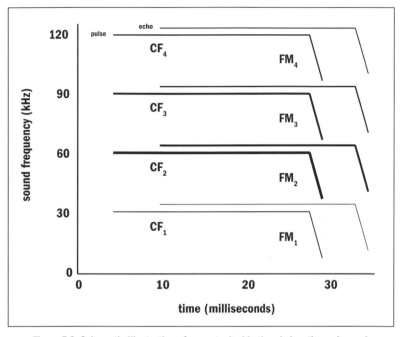

Figure 7.6. Schematic illustration of a mustached bat's echolocation pulse and its echo, in the form of a sonogram, which shows the frequencies of sounds over time. The thicker the line, the louder the sound. The bat's pulse (and its echo, which is delayed and is at a slightly higher frequency, due to its Doppler shift) contains a fundamental frequency (1) and three harmonics (2–4, each a multiple of the fundamental frequency). Each pulse is a constant-frequency tone (CF), followed by a frequency-modulated downward sweep (FM).

studies of the mechanisms of echolocation, much of it in the lab of Nobuo Suga at Washington University in St. Louis, so let's look at the research on this species more closely.[25]

The mustached bat emits sound pulses with a fundamental frequency that begins at about 30 kHz and includes three harmonics—multiples of the fundamental frequency—that begin at about 60, 90, and 120 kHz (see Figure 7.6). Each pulse consists of a relatively long constant-frequency

(CF) tone followed by a frequency-modulated (FM) downward sweep. The mustached bat suppresses much of the outgoing 30-kHz fundamental-frequency CF and FM, which still reach its ear through conduction along its bones, so the loudest component of the outgoing pulse (and the returning echo) is actually the 60-kHz second harmonic (CF_2).[26] The bat's brain has some sets of neurons that respond to the fundamental frequency—either CF sounds at about 30 kHz (CF_1) or FM downward sweeps (FM_1) that begin at about 30 kHz (from pulses)—and other sets of neurons that respond to harmonics (CF_{2-4} or FM_{2-4}, from the echoes). There are then combination-sensitive neurons that receive one input from neurons responding to the fundamental frequency (from the pulse) and another input from neurons responding to a harmonic (from echoes).

For pulse–echo delay, neurons that respond to the FM sounds are used, as they give more precise timing information. The routes through the brain to the combination-sensitive neurons create varying delays (as in the owl brain pathways for computing sound azimuth), so that each neuron receives coincident pulse and echo FM signals (and so spikes the most) only for a particular pulse–echo delay. Surprisingly, the synapses delivering the pulse and echo inputs each use a neurotransmitter that produces synaptic *inhibition,* not excitation, in the combination-sensitive inferior colliculus neurons.[27] The inhibition is apparently followed by precisely timed excitation, through ion channels that produce postinhibitory rebound, much as proposed by Graham Brown in his half-center hypothesis for rhythm generation (see Chapter 5).[28] In one part of the cerebral cortex, the neurons are anatomically arranged to form a map of pulse–echo delays (or object distances).[29] If a drug is injected to temporarily shut down this part of the cerebral cortex, then the bat can no longer tell the distance of an object (but can still tell frequencies apart).[30]

Bats are incredibly good at determining object distances using pulse–echo delays. In fact, some species trained to detect pulse–echo delays that alternate between a longer and shorter value ("jitter") can detect changes in delay as small as 10 nanoseconds (ten billionths of a second), which correspond to changes in object distance of about 3 micrometers, which is less than the diameter of a cell![31] This also explains how bats may use echolocation to determine the texture of an object: a rough object will create pulse–echo delays across its surface that differ from those created by a smooth object. This may help bats tell apart a moth and a leaf, for example.

Bats can also compare pulses and their echoes to determine how fast an object is moving toward or away from them. How does that work? Any time a sound source is moving toward an ear, the sound frequency increases; when the sound source is moving away, the sound frequency decreases. This increase and decrease of sound frequency is what you hear when a train or ambulance speeds toward you and then passes you and speeds away. This is called Doppler shift and is caused, essentially, by the sound waves being compressed as they move toward you and stretched as they move away from you. Bats usually fly while they're echolocating, so echoes even from stationary objects in front of them are Doppler shifted to a slightly higher frequency. So, to be precise, a bat does not detect the velocity of an object, but instead the relative velocity, or how quickly the bat and the object are approaching each other.

To determine relative velocity, the mustached bat relies on the CF components of the pulse and echo, which are optimal for detecting slight changes in frequency. The amount by which each harmonic is increased in frequency in the echo indicates the relative velocity of the bat and that object. There are enlarged areas of the cochlea, the inferior colliculus, and the cerebral cortex in this and other bat species in which neurons respond to the Doppler-shifted CF_2, so this frequency range is like an "auditory fovea," analogous to the visual fovea of many mammals and the "somatosensory fovea" of star-nosed moles.[32] There is a set of combination-sensitive neurons in the cerebral cortex (in a different area from that of the delay-tuned neurons) that respond selectively to the combination of an approximately 30-kHz CF_1 (from pulses) and a CF slightly above 60 kHz or 90 kHz (from the Doppler-shifted CF_2 or CF_3 of echoes), indicating a particular relative velocity.[33] Different neurons respond best to different frequency combinations and thus signal different relative velocities. Both the continuing relative velocity and fast changes in this velocity may be detected.[34] Such rapid frequency changes are caused by an object that alternately moves toward and away from the bat, such as the wings of a flying moth.

Given how efficient bats are at using echolocation to detect and track moths (see Figure 7.7), moths have long had an incentive to develop methods of counterintelligence. Some species of moths have evolved such methods. These species can hear ultrasound, unlike insects generally.[35] Some species react to ultrasound by flying erratically or diving to the ground. Other species actually emit their own ultrasound in response, at a high rate and intensity, apparently confusing the bat's interpretation of its echoes.[36]

Many species of echolocating bats live in large, dense colonies (in caves, for example) and hunt in large groups. This

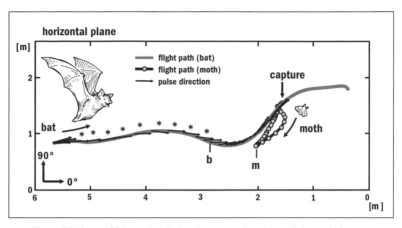

Figure 7.7. Actual flight paths of a bat (gray curve) and a moth (gray circles connected by black line segments) while the bat chased and captured the moth using its echolocation pulses. The direction in which the bat emitted each echolocation pulse is also shown (arrows along the bat's path). Asterisks indicate when the bat used a long constant-frequency (CF) pulse.

presents a potential problem for echolocation: How does a bat know if an echo it hears is from its own pulse or from its neighbor's pulse? Bats have evolved a number of mechanisms to reduce, if not eliminate, this problem. One mechanism is that when they are really densely packed, such as when flying into their home cave, they actually stop echolocating and rely on their memory of the route. This became apparent when a certain cave entrance in Indiana was suddenly blocked by a door. Soon after, thousands of bats were found dead just outside the entrance, having apparently collided with the door.[37] Presumably, they didn't notice the door was there, even though they are able to navigate precisely using echolocation, so they must have turned their echolocation off and flown on autopilot (or, more precisely, relied on their memory). Donald Griffin confirmed this finding in controlled laboratory tests by moving a barrier within a room previously familiar to the bats.[38] Some species also reduce their rate of pulse emission when they are hunting in close proximity with others of the same species ("radio silence"?), presumably to avoid "jamming" one another's sonar.[39] In other cases, however, bats purposely jam their neighbors' sonar when competing for insects.[40]

Electronic Surveillance

How animals avoid jamming one another's surveillance sensors has been more thoroughly explored in a completely different kind of animal: weakly electric fish (knifefish). These fish usually live and hunt in muddy water, often at night, so vision is not very useful. They generate an electric field from their tail that surrounds their body. This electric field is too weak to damage (or usually even be detected by) prey, but the weakly electric fish can precisely sense disruptions in this field, as Hans W. Lissmann discovered in 1951.[41] Animals are essentially composed mostly of salt water, which conducts electricity better than the fresh water that these fish live in. The greater conductivity of prey and predators alters the electric field the weakly electric fish has set up (see Figure 7.8). The fish detects this change via electroreceptors spread across its body surface.

Signals from the electroreceptors are sent to the brain, where they form a map of space in the tectum, as we saw for sounds and sights in barn owls and touch in the star-nosed mole. Some weakly electric fish species generate individual pulses at varying rates, analogous to the sound pulses of echolocating bats. Other species generate continuous electric waves (similar in shape to sine waves) at a particular frequency, analogous to a single-frequency sound.

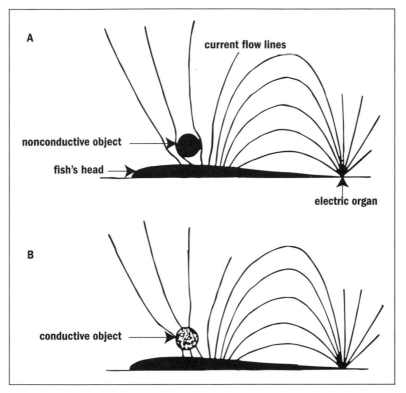

Figure 7.8. Lissman's depiction of how the electric field of a weakly electric fish is distorted by an object that is (*A*) less conductive or (*B*) more conductive than the water.

Weakly electric fish have a potential problem, however. They often swim near others of their own species that are generating very similar electric fields. There is a great potential for confusion where these fields overlap. To avoid such confusion, each fish alters the frequency of its own electric field to make it more different from the neighbor's field.[42] This behavior is called the jamming avoidance response (JAR; see Figure 7.9). The pathway through the brain that

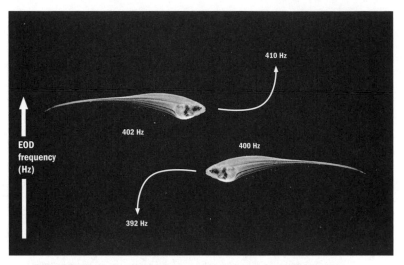

Figure 7.9. Illustration of the jamming avoidance response (JAR) of weakly electric fish, in which each fish changes the frequency of its electric organ discharge (EOD) to make it more different from its close neighbors. See color insert.

produces the JAR has been worked out in its entirety (see Figure 7.10) for a wave-type species, the glass knifefish *(Eigenmannia)*, mainly in the laboratory of Walter Heiligenberg. (Heiligenberg was unique: he came to work each day at the Scripps Institute of Oceanography in La Jolla, California, wearing a white turtleneck sweater, white shorts, and sandals. When he arrived on the morning of his fiftieth birthday, he found his lab members also wearing white turtlenecks and shorts, which his wife had secretly brought to them from his closet. Heiligenberg continued to work long hours performing his own experiments well into middle age, until his untimely death in a commercial plane crash in 1994.)

So how does the JAR work? If a neighboring fish is using a slightly higher or slightly lower frequency for its electric

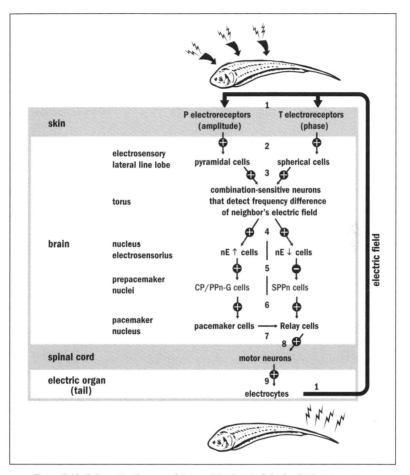

Figure 7.10. Schematic diagram of the weakly electric fish circuit that generates the jamming avoidance response (JAR). Electroreceptive sensory neurons in the skin (top of diagram) sense changes in the electric field in the water, activate combination-sensitive interneurons in the brain, and cause a change in motor neuron spiking in the spinal cord, which in turn alters the fish's electric organ discharge frequency (bottom of diagram). The pathway on the left increases the fish's electric wave frequency (when a neighboring fish is using a lower frequency), and the pathway on the right decreases the fish's frequency (when the neighbor is using a higher frequency). (CP/PPn-G = central posterior/prepacemaker nucleus-gradual; nE↑ = nucleus electrosensorius up; nE↓ = nucleus electrosensorius down; SPPn = sublemniscal prepacemaker nucleus.)

surveillance, the two frequencies add together in the water. This is analogous to two sounds at slightly different frequencies being played together. As any musician knows, when two instruments are slightly out of tune with each other and play the "same" note, their combined sound produces "beats" (a kind of slow "wah-wah" sound). The rate of these beats equals the difference in sound frequency of the two individual notes. Beats occur because the two sound waves alternately interfere with each other "constructively" and "destructively." In other words, the combined sound amplitude increases when the two waveforms are aligned, or in phase, and then decreases when the two waveforms are misaligned, or out of phase.

What happens with two slightly different frequencies of weakly electric fish electric waves is analogous, and the two fish monitor the resulting electric wave beats. Weakly electric fish detect electric fields using two different arrays of sensors in their skin (see Figure 7.10). One kind of sensor (P electroreceptors) detects the amplitude of each wave; the other (T electroreceptors) detects the precise timing of each wave peak. These two kinds of electric signals converge onto combination-sensitive neurons in the fish's version of the inferior colliculus (called the torus). The combination of the time and amplitude of the summed electric waveform tells the fish whether the neighboring fish is using a slightly higher or a slightly lower frequency. If the other fish is using a higher frequency, the combined waveform increases in amplitude whenever it is delayed with respect to the fish's own wave (phase delayed). If the other fish is using a lower frequency instead, the combined waveform increases in amplitude whenever it is ahead of the fish's own wave (phase advanced).

The combination-sensitive neurons in the torus each respond to signals from electroreceptors on one part of the skin. Some regions of the skin are dramatically affected by the other fish's signal, and other regions are barely affected at all (depending on where the neighboring fish is). Behaviorally, the fish relies on differences between the signals on different skin regions and cannot perform the JAR if the combined electric field is uniform throughout its body. Essentially, the fish compares a signal indicating the combined waveform (from skin regions that are dramatically affected by the neighboring fish) to a "reference" signal indicating its own waveform (from skin regions that are largely unaffected by the neighboring fish).

The brain performs the JAR using a "democracy" of combination-sensitive neurons to collectively determine whether the neighboring fish is producing a slower or faster signal (see Chapter 4). The neurons receiving signals from skin regions that are more affected by the other fish's electric wave spike more often and thus have a larger effect on the outcome. Some combination-sensitive neurons receiving signals from a weakly affected part of the skin actually "vote incorrectly" regarding the other fish's electric wave, but they have a low spike rate, so they are outvoted. This collective decision making is so efficient that it can detect changes in the timing of electric signals smaller than 1 microsecond (one millionth of a second)![43]

The signals from combination-sensitive neurons are sent through two brain pathways (see Figure 7.10), where they cause either an increase in the fish's own frequency (via the nE↑, the CP/PPn-G, and pacemaker cells) or a decrease in the fish's own frequency (via the nE↓, the SPPn, and relay cells). Interestingly, increased spiking of the SPPn cells actually causes an *increase* in the fish's frequency, but the SPPn cells' inputs are *inhibitory*—the stronger the inputs, the *less* active the SPPn cells are, so the effect of increasing activation of the nE↓ pathway is to decrease the electric wave frequency. The tallying of votes effectively occurs in the relay cells of the pacemaker nucleus, where the up and down pathways converge. The relay cells act via spinal cord motor neurons and then tail electrocytes, which are the cells that generate the electric field. The net result is to adjust the fish's own electric organ frequency, increasing it when the other fish is using a lower frequency and decreasing it when the other fish is using a higher frequency, which amplifies the difference between the two fish's surveillance frequencies and allows each fish to spy without electrical interference.

We can learn some lessons about surveillance from barn owls, star-nosed moles, echolocating bats, and weakly electric fish. First, there are multiple types of sensors that can be used to construct precise and panoramic images of the nearby world. Combining the signals from different types of sensors into a single "map" can increase the reliability of conclusions. In some cases, it is efficient to use one type of sensor to locate an object of interest and a second type of sensor to determine what the object actually is. Second, having a large number of neurons contribute to the analysis can lead to a more accurate and reliable conclusion than having just one neuron decide on its own. Third, systems must be in place to avoid interference from others using the same type of surveillance in the same vicinity.

8

Government Self-Monitoring

Governments are typically large and complex. There are multiple agencies, and it's difficult for one agency to know what another agency is doing. But if they don't know, they may duplicate their efforts or, worse yet, do things that are contradictory, disruptive, or incompatible. To avoid these conflicts, communication is crucial.

Similarly, nervous systems are large and complex and are divided into neuronal circuits that have different functions. These neuronal circuits must communicate with each other to coordinate their activities and avoid conflicts that could be disruptive. When neuronal circuits cause movements, for example, these movements often alter the sensory input that is received by other neuronal circuits. Thus, a circuit that causes movements might be expected to disrupt the functions of a circuit that senses the environment. This is a fundamental conflict that must be avoided at all costs. How can it be avoided?

Let's take a specific example. Imagine that you are playing baseball. You are an outfielder, running to catch a fly ball. The ball is heading behind you and to your left. You are running back diagonally to the left while keeping your eyes on the ball. What does your nervous system need to do so that the ball ends up in your mitt?

Obviously, you have to keep tracking the ball's location. This requires that your eyes are open and that the image of the ball remains near the center of your retina. But is this enough?

If the ball is centered on your retina and you are facing forward with your eyes looking straight ahead, then the ball is in front of you. But if the ball is centered on your retina and your eye or head is turned to the left, then the ball is to your left. If the ball is centered on your retina and your eye or head is turned to the right, then the ball is to your right.

So knowing that the ball is centered on your retina doesn't tell you where the ball is unless you also know which way your eyes and head are turned. You can't catch the ball unless you know these things. You've probably never thought about this, yet you are still able (most of you, anyway) to catch a fly ball. That means that your nervous system is taking care of this business even though you are unaware of it. Somehow, the neuronal circuits that produce visual perception must receive communications regarding your movements. How does your nervous system do that?

Hermann von Helmholtz thought a lot about this in the nineteenth century in Germany.[1] (He wasn't actually interested in baseball, but this issue is much bigger than baseball. We can't use our vision to do much of anything in the world unless we know which way our eyes are facing.)

Helmholtz suggested that there are three kinds of signals we might use to keep track of which way our eyes are turned within our heads. (We'll ignore for now how we keep track of which way our heads and torsos are turned. The issues are essentially the same, but it's simpler to focus on eye movements.)

Helmholtz described the first kind of signal as the "intensity of the effort of will," meaning how much we are trying to move our eyes to the left or right. His second kind of signal was the tension in the eye muscles (which we might detect using sensory neurons in the muscles). His third kind of signal was the "result of the effort," meaning how much our eyes actually move (which we might detect using sensory neurons in the skin around our eyes, for example). The second and third kinds of signals do not necessarily happen together. For example, sometimes we activate muscles to generate force but don't actually move any part of our body (this is what happens during isometric muscle contractions).

How can we tell which kind(s) of signal our brain is actually using to keep track of our eye movements?

Helmholtz devised clever experiments to answer this question. One of these you can actually try at home. Cover one eye with your hand. (Otherwise, you will see double images, which confuses the result.) With the tip of the index finger of your other hand, touch the skin just to the side of your open eye. Gently either push the skin (and your eyeball, via the skin) inward or pull it outward. While you are doing this, what is your visual perception of what the world in front of you is doing?

Most of us will find that when we move our eye to the left in this way, the world appears to move to the right. When we move our eye to the right, the world appears to move to the left. Why does it seem that way when the world is not actually moving?

This is actually not puzzling. When we move our eye to the right in this way, the visual image moves to the left across our eye. So of course we perceive that the world is moving to the left. That's what our eye is telling our brain.

The strange thing is that we *don't* perceive that the world in front of us is moving to the left when we make a voluntary rightward eye movement in the usual way. Why don't we?

Helmholtz reasoned that when we push or pull our eye with our finger, the eye is actually moving, yet our brain apparently doesn't take that eye movement into account, because we interpret the movement of the image on our retina as a real movement of the world. So our brain must not be keeping track of the result of the effort. If our brain is not keeping track of the result when we push or pull the eye, presumably it is also not using this kind of signal when we make eye movements in the usual way. By the process of elimination, that leaves intensity of the effort of will and muscle tension.

So Helmholtz did another experiment. (Please *don't* try this one at home.) There are patients who have paralysis of one eye muscle alone, for example, the one that turns the right eye to the right. When such a patient covers their left eye only and then tries to look to the right, the patient reports that the world in front of them has jumped to the right. (In reality, not only has the world not jumped, but their right eye also has not moved.)

If neither the world nor the patient's right eye has moved, why does the patient perceive that the world has moved to the right? Helmholtz suggested that the patient's brain *assumed* that the right eye really did move to the right. If the eye really had moved, and if the visual image was still in exactly the same location on the retina after this movement, then the only way this could have happened is if the world moved to the right by exactly the same amount as the eye moved. So that's what the brain concluded and that's what the patient perceived.

Why would the patient's brain assume that the right eye had moved when it did not move? The eye muscle was paralyzed, so there was no tension in the muscle and no actual eye movement. The only possible signal remaining was the intensity of the effort of will. In other words, the patient's brain assumed that the eye moved to the right simply because the patient *tried* to move their eye to the right. The brain knew what movement was intended and assumed that the movement command had been carried out. So we can perceive the real locations of objects we see (like a fly ball) because our brain keeps track of its own movement commands.

This was a brilliant hypothesis! In fact, Helmholtz (like Graham Brown) was way ahead of his time—in Helmholtz's case, by almost a century.

Jump ahead to 1950. For some reason, the time was just right. Scientists in two countries—Roger Sperry in the United States, and Erich von Holst and Horst Mittelstaedt in Germany—independently returned to Helmholtz's idea and tested it in animals.[2]

Sperry studied fish. He anesthetized a fish and then surgically rotated one of its eyes so that it was upside down. Then he covered the normal eye with an opaque cap and allowed the fish to recover from the anesthesia. What he found is that the fish swam in tight circles after this operation. Whichever way it began circling (to the left or the right), it would continue circling that way.

Von Holst and Mittelstaedt designed a very similar experiment (independently from Sperry), but they used a type of fly with a long neck. They rotated the neck so that the fly's entire head was upside down and glued the head in that upside-down position. These flies also flew in circles.

Both teams reasoned that the left–right visual reversal they had created by their manipulation must have somehow caused the circling. What each supposed, following Helmholtz, was that normally, when the animal turns to the left, the image on its eye (if the world is not moving) moves to the right. Somehow, the animal's brain expects that this will happen and compensates for it. But with the eye or head upside down, when the animal turns to the left, the image on its eye actually moves to the left, instead of to the right. This is not what the animal's brain expects, so it does not compensate appropriately.

Let's take a specific example. If the animal turns 30 degrees to the left, normally the visual image moves 30 degrees to the right on its eye. (In a vertebrate eye, the image actually moves to the *left* on its *retina*, because the eye's lens reverses the image. But we'll ignore the image reversal normally performed by the eye, because it doesn't matter for our discussion and would be unnecessarily confusing.) If that actually happens, the brain "concludes" from the data available to it that the world has not moved at all, so that is the animal's perception. But if the animal turns 30 degrees to the left and the image on its eye moves 30 degrees to the left (when the brain expected it to move 30 degrees to the *right*), the brain concludes from the available data that the world must have moved *60 degrees to the left,* as that is the only way those two events could have occurred together.

Most animals like to keep their eyes focused on some point in the visual world and not have their visual world appear to jump around. So if the world has just moved 60 degrees to the left, the animal turns 60 degrees to the left to catch up with it. But when the animal (with upside-down eye or head) makes this turn, the image on the eye actually moves still farther to the left (another 60 degrees now). So the animal turns still farther to the left. And now the animal is circling.

This whole interpretation of why the animal is circling only makes sense if the animal's brain expects a certain movement of the image on the eye and somehow compensates for the expected movement. Sperry hypothesized that along with the movement command to turn the eyes or head or body, the brain produces a "corollary discharge," meaning spiking in neurons that goes along with the movement command (like a corollary goes along with a theorem in math), which is sent to neuronal circuits involved in perception. Von Holst and Mittelstaedt hypothesized that the brain issues an exact copy of the movement command, which they called an efference copy. ("Efference" just means outgoing signals—toward the muscles—while "afference" means incoming signals—from the senses.)

Whether we call this signal corollary discharge or efference copy or intensity of the effort of will, it amounts to basically the same thing: the brain keeps track of its movement commands and somehow compensates for them in perception. (I will mostly use "corollary discharge," which is a somewhat broader term than "efference copy," because it does not assume that the signal our brain uses to keep track of the intended movement is an exact copy of the movement command signal. "Intensity of the effort of will" is just too cumbersome.)

That is, our brain assumes that the movement we intended to carry out was actually carried out and that the visual image on the eye (and, potentially, sensory inputs from other senses) changed accordingly. If the expected change in sensory input occurs, our brain somehow "cancels it out" and we perceive that the world is stable. If the actual sensory input is different from what we expected, however, then we perceive that something in the world has really moved. In essence, the neuronal circuits responsible for movement inform the neuronal circuits responsible for visual perception of their plan of action.

Von Holst and Mittelstaedt defined some additional terms: "reafference" to refer to sensory inputs that are *caused by* our own movements (such as the movement of an image across the eye due to a voluntary eye movement), and "exafference" to refer to sensory inputs that are caused by a change in the external world, not by our own movements. All sensory inputs taken together (reafference + exafference) are called simply "afference."

Von Holst and Mittelstaedt imagined that the brain uses the efference copy signal in a quantitative, quasi-mathematical way. They hypothesized that the part of the total afference caused by our own movements (reafference) is canceled out by the efference copy, leaving behind in the brain the remaining afference (exafference) to be perceived as an actual change in the world. They envisioned this process as akin to arithmetic addition (see Figure 8.1).

For example, let's arbitrarily define rightward as positive and leftward as negative. If you keep your eyes (and head and body) still and something in the world moves 30 degrees to the right (see Figure 8.1A), there is no efference or efference copy signal (because you have not tried to move your eyes). There is, however, an afference signal of +30°. The afference signal and the efference copy are added together somewhere in the brain: $+30° + 0° = +30°$. So you perceive that the object (X in Figure 8.1) has moved 30 degrees to the right.

If, instead, you use your finger to move your eye 30 degrees to the right (as you may have done earlier, following Helmholtz's directions), the efference copy signal is 0° (because you have not issued a normal eye movement command), the afference is –30° (because the image did move to the left across the eye), and the sum of the two is $0° - 30° = -30°$ (see Figure 8.1B). So you perceive that the world moved 30

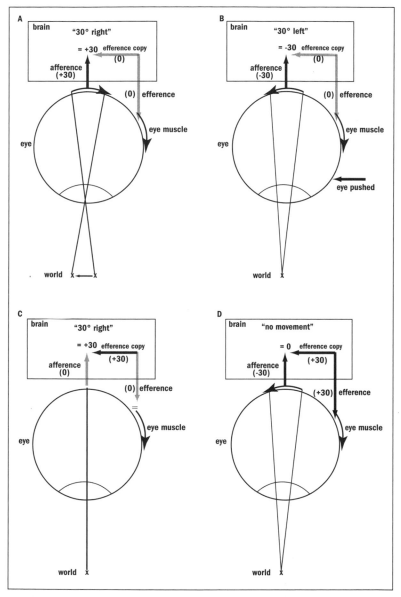

Figure 8.1. Diagram of von Holst and Mittelstaedt's hypothesis for eye movements. See text for details.

degrees to the left (as you probably experienced when you did this experiment on yourself).

If, on the other hand, the muscle that moves your right eye rightward is paralyzed (as it was for some of Helmholtz's patients) and you try to move your eye 30 degrees to the right, then the afference is $0°$ (because your eye did not actually move), the efference copy is $+30°$, and the sum is $0° + 30° = +30°$ (see Figure 8.1C). So you perceive that the world jumped to the right, as Helmholtz's patients informed him.

Finally, if you make a voluntary eye movement 30 degrees to the right (in the usual way), there is an efference copy signal of $+30°$ and the image of the (stable) world moves 30 degrees to the left on the eye, so the afference is $-30°$. The efference copy and afference signals cancel out: $+30° - 30° = 0°$. This tells the brain that the world movement is zero; in other words, the world has not moved.

In a sense, then, what happens during a normal eye movement (Figure 8.1D) is the simultaneous occurrence of the eyeball-push experiment (Figure 8.1B) and the paralyzed eye experiment (Figure 8.1C).

So von Holst and Mittelstaedt's quantitative approach works well to account for the results of these experiments.

This all sounds very reasonable, but it would be nice to be able to actually detect these efference copy or corollary discharge signals directly, instead of just assuming they exist because of the resulting behavior and perception. In other words, instead of treating the brain as a "black box" whose contents we must infer from behavior and self-reported perception, can we open up the black box and look at the relevant signals directly?

Marc Sommer and Robert Wurtz opened up the black box by placing microelectrodes into the brains of trained monkeys.[3] They trained the monkeys to stare at a spot of light on a screen (the fixation spot) while sitting in a dark room.

Then, two "target" lights flashed sequentially at locations off to the side. Next, the fixation spot went off and the monkey had to make one eye movement (a saccade, as we discussed in Chapter 4) to the first target light and then a second saccade to the second target light. The monkey had to remember where the two light flashes had been. The only way the monkey could make the second saccade accurately was if its brain took into account that it had already made the first saccade. In other words, the brain had to be updated on which direction the eyes were facing (in the dark) *after* the first saccade in order to make the second saccade to the correct location. One way for the brain to keep track of the saccades would be corollary discharge signals. The monkeys learned to do this task very well.

Then Sommer and Wurtz used a microelectrode to monitor neurons (in a part of the brain called the thalamus) that spiked just before the monkey made an eye movement. Sommer and Wurtz demonstrated that the axons of these neurons went to the cerebral cortex, not toward the eye muscle motor neurons (in the brainstem), which suggested that these neurons were *not* causing eye movements. Sommer and Wurtz hypothesized that these thalamic neurons were instead carrying a corollary discharge signal.

Then Sommer and Wurtz injected a drug into this part of the thalamus to inhibit the hypothesized corollary discharge neurons. After they injected the drug, when the monkey repeated the task, the first saccade in each pair of saccades was accurate. The second saccade, however, was off target. The second saccade was shifted in a direction consistent with the monkey's brain *not* taking into account that it had made the first saccade.

This suggested that the monkey's brain was indeed using a corollary discharge signal that had been disrupted by the drug.

This monkey experiment showed that corollary discharge signals can be identified within a primate brain, if not yet a human brain. But we cannot yet work out the entire neuronal circuit for visual perception in a monkey and determine exactly how the corollary discharge signal interacts with visual perception circuits. To get into this level of circuit detail, we really need a more suitable animal.

How about a cricket?

Male crickets chirp (stridulate) to attract females of their species.[4] This is an important task. As we've seen, crickets even have a command neuron to trigger this behavior (see Chapter 3).

One potential problem, however, is that this chirping is very loud (as loud as 100 decibels near the animal). Loud is good if you're trying to attract mates from far away. Loud is not good if you want to keep hearing things. (It's also not good for people nearby who are trying to sleep.) In particular, you might want to hear a predator approaching even while you are making such a racket. But the loud sounds you are making might saturate your auditory system so that you can't detect other sounds during or just after these loud sounds. How can you have it both ways?

In principle, you could have it both ways if you could "cancel out" the effect of your own chirping on your own auditory system, something like noise-canceling headphones. The best way to do that would probably be to keep track of the timing of your own chirps, using a corollary discharge signal, and then somehow counteract the effect of the chirps on your own hearing. If you counteracted the effect of your own chirps by just the right amount, your nervous system would not respond to the chirps but might still respond to additional sounds. But do crickets actually do this?

James Poulet and Berthold Hedwig showed that they do.[5]

Poulet and Hedwig discovered a neuron in crickets that they named corollary discharge interneuron (CDI). (Unfortunately, neurophysiologists—unlike fruit fly geneticists—typically don't invent entertaining names for their discoveries!) They poked a microelectrode into CDI and monitored its spiking (Figure 8.2).

CDI spiked during chirping (Figure 8.2A). So it might be causing the chirping, or it might be responding to the sound of the chirping. But causing CDI to spike at a different time (by injecting positive current into CDI) did *not* alter chirping (Figure 8.2B), so it is unlikely that CDI causes the wing movements that underlie chirping. Also, CDI did *not* respond to the sound of chirping if this cricket was not the one making the sound (Figure 8.2C), so CDI also appears not to be involved in hearing chirps. This means that CDI is *not* a movement control neuron *or* an auditory neuron, even though it spikes during chirping. So CDI might carry a corollary discharge signal instead.

To find out, Poulet and Hedwig poked a second microelectrode into a known auditory interneuron named ON1. They saw that ON1 is inhibited (its voltage goes down) whenever CDI spikes during chirping (Figure 8.2D). When they caused CDI to spike at a different time (by injecting positive current into CDI), ON1 was also inhibited by this (Figure 8.2E).

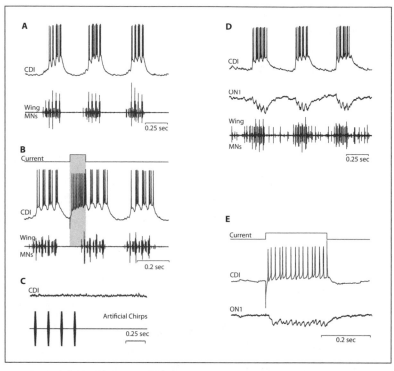

Figure 8.2. The cricket corollary discharge interneuron (CDI) during chirping. *A*. CDI spikes with chirps. *B*. Injecting positive current into CDI to make it spike at another time does *not* alter chirps, so CDI is not affecting chirping movements. *C*. CDI also does not respond to the sound of (artificial) chirps. *D*. During chirping, CDI inhibits an auditory interneuron, omega neuron 1 (ON1; its voltage goes down when CDI spikes). *E*. Injecting positive current into CDI when there is no chirping also inhibits ON1.

So CDI reduces the auditory response in the brain (by inhibiting ON1—as well as by inhibiting sensory neurons in the ear, but those data aren't displayed here) during each chirp. In other words, CDI cancels out the effect on the brain of each chirp sound, much as the efference copy cancels out reafference in von Holst and Mittelstaedt's scheme. If the cancellation is of just the right size, this should still allow other (environmental) sounds to be heard by this cricket even while it's making such a racket.

So we can identify neurons that carry a corollary discharge signal in animals as diverse as monkeys and crickets. Corollary discharge signals have also been shown to occur during many other animal behaviors, including mollusk feeding, crayfish escape, crayfish walking, cockroach walking, locust head movements, electric fish electric organ discharges, vocal fish calling, skate (a fish) ventilation, tadpole swimming, zebra finch singing, rat whisker movements, cat walking and scratching, human walking, and human hand movements, among others.[6] Corollary discharge signals during movements can modify visual, auditory, and touch (the somatosensory system) perception, as well as the sense of balance (the vestibular system).

As we've seen, command neurons (discussed in Chapter 3) can inhibit sensory inputs in the near future that are likely the result of the movements the command neuron triggers— this is what happens when a spike in the crayfish LG neuron initiates inhibition of subsequent abdominal sensory inputs. The LG spike also causes inhibition of extensor muscle stretch receptors that would otherwise spike as a result of the strong flexion during an LG tail flip. So both of these pathways also effectively convey corollary discharge signals. In this case, the command neuron signals multiple neuronal systems of its decision and makes sure that these other systems don't do anything that would interfere with carrying out the command.

Central pattern generators (discussed in Chapter 5) in animals from crayfish to cats also cause rhythmic inhibition of inputs from sensory neurons and sensory interneurons. Because of this rhythmic inhibition, sensory inputs received at a particular time within a cycle are less likely to elicit a response—for example, you would not want to withdraw your leg (unless absolutely necessary) at a time when you are relying on it to support your weight—and this is what happens in locomoting animals generally. These are also corollary discharge signals. In essence, the neuronal system responsible for running a factory regulates feedback signals that normally alter its operation, so that the feedback won't have too large an effect at an inopportune time.

Although the number of confirmed examples of corollary discharge neurons is still relatively small, it is very likely that all neuronal circuits required for either sensory perception or movement control take into account the animal's own movements and do so at least partly via corollary discharge neurons. These circuits allow animals to latch onto important events in the environment, such as the presence of a predator, prey, or mate, while not being distracted or confused by sensory inputs caused by their own movements. Without such self-monitoring of decisions and plans, our neuronal circuits would constantly disrupt and interfere with one another. So corollary discharge is a key method of communication within our nervous systems.

And it also allows you to catch a fly ball.

9

Becoming a Political Animal

Our lives involve a lot of politics. We each try to acquire allies and deter rivals; our governments do similar things on a larger scale. This often involves trying to appear strong and reliable.

We begin learning how to interact politically at an early age. (Think of preschool.) Arguably, much of the learning that occurs in social animals like humans is learning how to manage social interactions with other humans. The nervous system, of course, is what mediates this kind (and other kinds) of learning.

We cannot investigate at a cellular level how such learning occurs in people. But many animals face similar challenges: they also have to gain allies and deter rivals. They do so in many ways, both simple and complex. For example, many animals urinate to mark their territories and ward off rivals. Some carnivores make loud sounds (such as a lion's roar) to announce their presence. In many species, males display themselves to attract mates and ward off rivals. Males that appear larger and more attractive are more likely to succeed at both. Being more attractive is usually correlated with being healthier.

Some of the best animals for studying how nervous systems manage social interactions are songbirds. Like humans, songbirds use both visual displays and vocal communication to argue their cases before potential allies and rivals. The songs that male birds sing, like their plumage, indicate their state of health and serve to announce their territory, attract mates and allies, and deter rivals. (Among birds, more colorful plumage often indicates greater health and energy storage.)

Using vocal communication to gain allies and dissuade rivals is a sophisticated approach that has much in common with human social and political behaviors. Singing by songbirds serves purposes similar to those found in diverse human verbal exchanges, from trash-talking among high school basketball players to formal debating among presidential candidates. In both birds and humans (but few other animals), success or failure in important aspects of life depends on the skill with which they conduct vocal exchanges. Birds, like humans, gradually learn to perform well in these kinds of contests. Bird singing is analogous to human speech, and birdsong learning is analogous to human language development. Although we cannot study at a cellular level how the nervous system develops its control of language in humans, we can do so in birds.

For these reasons, birdsong singing and song learning have long been favorite topics for those who study how nervous systems govern behaviors. The nervous system control of birdsong is complex and incorporates several features we have previously encountered, including combination-sensitive neurons, central pattern generators, specialized neuronal systems, and corollary discharge.[1] In addition, it exemplifies how nervous systems change to mediate behavioral learning. Birdsong is thus a suitable last example for us to explore in considering how nervous systems govern behaviors.

For many songbirds, the songs they sing are too complex to be encoded directly into their genomes, so they must be learned. The nervous system of each songbird species both facilitates and constrains this learning. For example, white-crowned sparrows can learn a version of another species' song if they never hear their own, but will choose to learn their own species' song (and will learn it better) if it is among those they hear.[2]

In most species, it is mainly the males that sing. A juvenile male typically learns to sing a song similar to one it has heard frequently, which is usually the song of its father. There is a particular period of juvenile development during which it needs to hear such a "tutor song."[3] If it does not hear its species' song during this sensitive period, it will never sing a proper song. This is analogous to what happens in human language development, which includes a sensitive period for hearing spoken language.[4] This is why people who move to a new country and begin speaking a new language late in childhood or as an adult generally speak with an accent, even many years later. Birds, too, have accents that match the way songs are sung in their locale.[5]

Juvenile birds typically begin memorizing a tutor song before they begin singing. Depending on the species, they may have to complete song memorization before they ever try to sing. Then, they try to imitate the song they memorized. At first, the noise they make is completely unlike what they heard (as with the babbling of human babies). Gradually, they produce closer and closer mimics of the tutor song. If all goes well, they eventually produce a song that is very similar but not identical to the tutor song. They incorporate some of their own idiosyncrasies, so that each song is unique to the singer and helps identify him.

A juvenile bird has to hear itself sing in order to refine its song. This suggests that the bird must compare the sound it makes to the sound it remembers and then make adjustments, perhaps through trial and error, to generate a better match, as Mark Konishi first suggested.[6] If a juvenile bird is deafened after it has memorized a tutor song but before it has begun singing, it never sings a normal song. (This is also parallel to human language learning, which is permanently affected by being deaf during the sensitive period for learning to speak.)

Birds use largely separate neuronal systems to sing and to learn to sing. Much of the research on how the brain controls singing and song learning has been done on zebra finches (Figure 9.1). The system that generates singing is organized hierarchically (see Figure 9.2, roman font). At the top of the hierarchy is a forebrain structure called HVC. Some HVC neurons project their axons to the brain structure just below them in the hierarchy, RA. RA neurons, in turn, project axons to motor neurons, which themselves excite skeletal muscles of the syrinx, in the throat. The syrinx, like the mammalian larynx, vibrates to generate a wide variety of sound frequencies and timbres as air flows by.

How do we know that HVC is at the top of the singing hierarchy? Birdsongs themselves are organized hierarchically in many species: distinct brief sounds (each called a syllable) occur in a particular sequence, with some syllables repeated a certain number of times, to form a complex but consistent song "motif" that the bird repeats. HVC neurons that keep their axons within HVC spike throughout the song motif (but not at other times).[7] Electrical stimulation of HVC with a microelectrode disrupts the entire motif, sometimes causing the song to end immediately and restart from the beginning.[8] In other cases, electrical stimulation of HVC ends the current syllable and restarts the song motif at that

Figure 9.1. A breeding pair of zebra finches *(Taeniopygia guttata)*. The male is on the left. Bird Kingdom, Niagara Falls, Ontario, Canada. See color insert.

point, either starting that syllable over or skipping it; the rest of the syllables are delayed or advanced accordingly. These effects indicate that HVC controls the song as a whole, like an upper administrator. If one inserts a thermal probe into HVC to cool those neurons gradually, slowing down their action potentials and synapses, then the bird continues singing the same song, but more slowly.[9] This suggests that HVC contains a central pattern generator for singing the song motif (which must be much more complex, however, than the central pattern generators that produce simple rhythmic behaviors like crustacean chewing and tadpole swimming). If HVC is destroyed, the bird cannot sing.[10]

Similar experiments on RA suggest that it occupies a lower level of this hierarchy. Each RA neuron spikes during

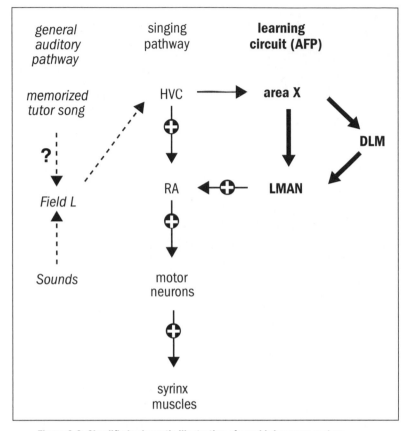

Figure 9.2. Simplified schematic illustration of songbird nervous system pathways to hear sounds (text in *italics* with dashed arrows), sing songs (roman font and solid arrows), and learn to sing (anterior forebrain pathway [AFP]; bold font and thick arrows). (DLM = medial nucleus of the dorsolateral thalamus; LMAN = lateral magnocellular nucleus of the anterior neostriatum; RA = robust nucleus of the archistriatum.)

a few brief moments within a bird's seconds-long motif, such as just before particular syllables.[11] Electrical stimulation in RA often perturbs one syllable, but does not stop and restart the entire motif and does not alter the timing of the remaining syllables.[12] HVC neurons that project their axons

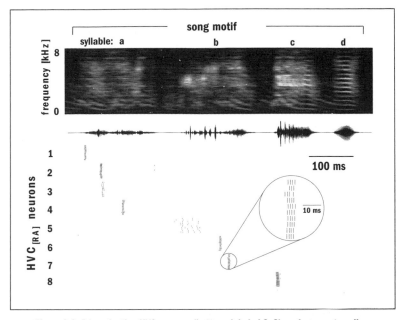

Figure 9.3. RA-projecting HVC neurons (bottom, labeled 2–8) each generate spikes (colored vertical lines) briefly and consistently (multiple trials are stacked) at one particular moment within the bird's song motif (top). See color insert.

to RA, which are excitatory (unlike the HVC neurons that keep their axons within HVC, which are inhibitory), spike at a brief and consistent moment within each seconds-long song rendition, often varying the time of their spikes by less than 10 milliseconds (Figure 9.3)![13] In at least some cases, an RA-projecting HVC neuron spikes just before the RA neuron spikes. RA neurons thus appear to act as "middle managers," relaying orders from HVC neurons to motor neurons, probably triggering the production of one syllable.

Another indication that HVC neurons sit atop a hierarchy is that some HVC neurons increase their rate of spiking selectively when the bird hears its own recorded song motif played back.[14] If only one syllable of this song is played, such a neuron does not spike much, but if two syllables are played in the correct sequence, it spikes a lot.[15] If the same sounds are played backward, the neuron barely spikes at all. Thus, these neurons act like complex combination-sensitive neurons or grandmother cells, responding selectively to the bird's own song (or large chunks of it). A more extreme version of such grandmother cells occurs in a particular species of wren that normally sings duets with a member of the opposite sex, with the two birds rapidly alternating syllables. In these birds (both males and females), some HVC neurons spike a lot only when both the male and the female syllables are played back to the bird in the appropriate order.[16]

The same HVC neurons that selectively increase spiking when a bird's own song is played also increase spiking when the bird itself sings.[17] This singing-related spiking occurs slightly earlier in the song motif and so appears not to be a response to hearing what it has just sung. Also, the playing of "noise" (such as another song) disrupts such a neuron's spiking in response to a heard song but not its spiking when the bird itself sings.[18] These findings suggest that these HVC neurons receive two kinds of inputs: one is a set of signals that indicates a particular combination of sounds, and the other is either a singing command signal or a singing corollary discharge signal. It has been suggested that these neurons may be like "mirror neurons," which are neurons found in monkeys that spike whenever the monkey makes a certain movement *or* sees another monkey or a human making the same movement.[19] Some researchers have suggested that such mirror neurons in humans (and perhaps in other animals) mediate empathy.[20]

HVC neurons other than those that project their axons within HVC or to RA project their axons to Area X, which is a part of the anterior forebrain pathway (AFP; see Figure 9.2, bold font). Connections among AFP structures form a loop: Area X neurons have axons that project to DLM; DLM neurons have axons that project to LMAN; LMAN neurons have axons that project to Area X. Other LMAN neurons have axons that project to RA, so the AFP as a whole receives input from HVC in the singing pathway and sends output to RA, also in the singing pathway. The AFP is thus connected in a way that could allow it to modify the singing pathway. The AFP is not necessary for singing in adult birds: destruction of one of its structures has no dramatic, short-term effect on singing.[21] But the AFP is required for song learning: destruction of one of its structures in a juvenile bird prevents the bird from learning to sing a normal song. Thus, the AFP is a separate brain circuit, specialized for learning to sing, but interacting with the circuit that is specialized for singing. Some neurons in the AFP, like some HVC neurons, increase spiking selectively in response to hearing the bird's own song, so they could be involved in listening for the "correct" song output.[22]

Adult songbirds can learn to modify their song, though juveniles learn faster, much as we saw for owl sound localization. For example, if adult birds are deafened after they've learned to sing normally, they continue to sing normally at first, but then their song gradually degrades over a period of months.[23] In addition, if a microphone and bird headphones are used to play back to an adult bird the song it has just sung, but with a slight delay (similar to what happens if you call in to a radio station and keep listening to the radio during your on-air call), then the bird's song also degrades over time.[24] This suggests that adult birds continue to compare what they intend to sing with the sounds they actually sang and try to make adjustments. However, Michael Brainard and Allison Doupe, at the University of California, San Francisco, showed that if an adult bird is deafened *and* a part of its AFP is destroyed, the song remains the same![25] This suggests that the AFP is necessary for *changing* the song (but not for singing itself), both in juveniles and in adults, and triggers changes even when changes are not helpful (as in the case of deafened adult birds).

So how does the AFP mediate learning to sing? This is probably more complicated than any of the other questions we have examined, and the process is not yet well understood. But intriguing clues have emerged and continue to emerge each year. Let's look at some of these clues.

First, LMAN generates some of the *variability* of a song. If LMAN is destroyed or its spiking is eliminated by a drug injection, the bird actually sings a *more consistent* motif than it normally does.[26] This implies that LMAN neuronal activity normally makes birds sing *less consistently*. Why would birds want to do this? The most likely hypothesis is that song variability is necessary for song learning. Trial-and-error learning can only work if the bird doesn't sing the same song every time. The bird has to try out different versions of the song and see what works best (perhaps evaluated on the basis of mate attraction or rival deterrence). Experimentation is necessary for learning. LMAN generates some of this variability. (Interestingly, when a juvenile male has a female in front of him, he immediately sings his song more consistently.[27] Apparently, that is not the time for experimentation—that is the time to deliver the best pick-up line he currently has and take his shot!)

Second, LMAN spiking is necessary for short-term memory storage in song learning. Brainard's lab developed a very precise way to train birds to sing a slightly different song.[28] Each bird is outfitted with headphones and a micro-phone. The microphone picks up the bird's song whenever it sings it. Whenever a certain syllable in the motif is reached, a computer detects it. If this syllable is sung below a certain threshold pitch, the software triggers either a brief noise or a played-back syllable that is slightly lower than the syllable that was actually sung. In either case, the bird responds to this unexpected feedback (perhaps thinking, did I really just say that?) by singing that syllable a little higher the next time around, to avoid the annoying noise or too-low pitch. The bird continues to raise the pitch of this one syllable as long as the experimenters play this game (with the threshold pitch raised by the experimenters each day to keep this process going). The experimenters can also play this game in one di-rection for several days, to cause the bird to sing this one syllable at a gradually higher pitch, then switch to playing the game in the opposite direction (either playing noise or playing back a syllable that is artificially elevated in pitch whenever the bird sings that syllable above a cut-off pitch) to cause the bird to sing the syllable at a gradually lower pitch. The experimenters can thus precisely control and measure learning of this one song syllable.

Aaron Andalman and Michale Fee, at the Massachusetts Institute of Technology, injected LMAN with a drug to elim-inate its spiking.[29] They did so at the end of one day, fol-lowing several days of this kind of training that altered the pitch of one song syllable. Each bird immediately sang the targeted syllable at the pitch it had used at the beginning of that day. Thus, without LMAN spiking, the bird loses all its learning from that one day, but retains the learning from pre-vious days. This shows that LMAN spiking stores same-day song learning of pitch, while learning of pitch from previous days is stored elsewhere in the brain (off-site storage?).

Third, although LMAN generates some of the song variability, birds can still learn pitch changes in the absence of LMAN's contribution to variability. This was demonstrated by blocking LMAN's outputs to RA, through injection of a drug into RA that selectively blocks LMAN's neurotransmitter receptors there, without blocking LMAN spiking.[30] This manipulation prevented LMAN from having an effect on RA (and thus prevented LMAN from increasing song variability) but did not stop LMAN neurons from doing other things (such as having effects within LMAN). During the injection, the birds showed no change in syllable pitch during a day of training. As soon as the drug injection ended, however, the birds immediately sang the targeted syllable at an appropriately higher or lower pitch. This means that learning actually did occur, but was not observable until the drug treatment ended. It also means that LMAN outputs to RA and the resulting song variability are not necessary for pitch learning. There must be other sources of song variability that are sufficient for this kind of learning. But LMAN spiking is still necessary for pitch learning. So LMAN has a role in pitch learning that is separable from generating variability.

Fourth, Bence Ölveczky's lab at Harvard University showed that although the AFP is important for pitch learning, a different circuit mediates learning of syllable timing.[31] When birds were trained to alter either the pitch of a particular syllable or the delay between a particular syllable and the next syllable (using an analogous kind of training), destruction of Area X (in the AFP) prevented the pitch change but not the syllable timing change. So what caused the syllable timing change? The experimenters used a microelectrode to monitor spiking in HVC neurons. The HVC neurons altered their spiking with the change in syllable timing, but not with the change in syllable pitch. This suggests that HVC is part of the circuit through which syllable timing, but not syllable pitch, is altered. Also, groups of projection neurons are clustered within HVC according to the time that they spike within the song when birds sing.[32] Although there is still much to be worked out for both kinds of learning, this suggests that a bird's brain has delegated different aspects of song learning—pitch and timing—to different brain circuits.

You may still be wondering, how does the bird compare the song it sings with the song it expects to hear, based on its memory of the tutor song? The answer is still uncertain. One hypothesis is that there are single "comparator" neurons that receive one set of inputs from the heard song when this bird sings it, and another set of inputs from its memory of the tutor song. Such neurons might spike selectively either when the two inputs match or when they don't match. But is there any evidence that such neurons exist?

For many years, no evidence was found. Then, such neurons were discovered in a surprising place: not in the specialized singing or song-learning circuits, but instead in the general auditory structures that receive signals from all kinds of sounds (see Figure 9.2, italicized font).[33] The experimenters used the headphones-and-microphone approach to play back an occasional wrong note while the bird sang and listened to its song. Certain neurons in these brain regions spiked only when there was a wrong note played and only if this happened while the bird was singing (not while it was just listening).

These "perturbation-selective" neurons appear to have all the qualifications for a job in which they detect singing errors. But how their inputs detect errors and how their outputs lead to corrections (if they do) remain to be worked out. One simple hypothesis is that they receive one input carrying auditory signals from the heard song (but only while they're singing) and a second input carrying corollary discharge signals that represent the singing command or intended song. If one of these inputs is excitatory and the other inhibitory, they might cancel each other out and prevent spiking, unless there is a mismatch between the intended and heard songs, in which case the neuron would spike at the moment of the mismatch, announcing a mistake. Such a neuron would be a new kind of combination-sensitive neuron.

There is still a great deal that we do not understand about songbird singing and song learning, due to the complexity of the behaviors and the complexity of songbirds' brains. But experiments in which spikes in brain neurons are monitored or manipulated while birds sing or learn to sing continue to reveal new surprises each year. Much of what we find out about these species may apply to what happens in our own brains when we speak or learn to speak. So the lessons from research on songbird neuronal circuits may shed light on the proximate mechanisms that underlie some of our most distinctively human abilities.

10

Governing Behavior

For most countries, it's relatively easy to tell which form of government they have.

Some governments are clearly dictatorships, with one well-known figure publicly issuing decisions and proclamations. The dictator may go on television to make speeches and be celebrated in posters and advertisements as the supreme leader.

Other governments are clearly a form of democracy, such as a representative democracy or republic. These countries often have a single, overarching document (such as a constitution) that outlines how the country's laws are supposed to function.

As we've seen, nervous systems govern behaviors in ways that are broadly analogous to human governments. Some behaviors can be triggered by a single dictator neuron. For example, a single spike in the lateral giant neuron in crayfish or the Mauthner neuron in fish triggers an escape. Other behaviors are generated by the collective voting of a large population of neurons, such as the jamming avoidance response in weakly electric fish and saccadic eye movements in monkeys. Nervous systems use central government programs to generate the basic, rhythmic outputs that keep animals alive, such as chewing, peristalsis, breathing, and locomotion. Nervous systems, like governments, perform surveillance of the environment to detect dangers and opportunities.

Neuronal circuits communicate with one another to keep track of one another's outputs, much as government agencies communicate their plans with one another (when all goes well).

But there are also important differences between the governments of animals and those of countries. One difference is that nervous systems don't follow a single playbook that is clearly laid out like a constitution and set of laws. Nervous systems have evolved over long periods of time to govern behaviors in whichever ways are effective, building on whatever previous forms of government existed. As a result, nervous systems use multiple playbooks, often at the same time. They can embody distinct forms of government simultaneously, even for control of a single behavior.

Thus, we have seen that for fish and crayfish escape movements, a dictatorship coexists with an oligarchy. Both may contribute simultaneously to the normal escape behavior. And if the dictator neuron is killed, the oligarchy carries on, without needing time to reorganize the government.

Similarly, the rhythmic movements that keep animals alive—such as breathing and locomotion—are not produced by a central program alone or by feedback mechanisms alone. Instead, they are produced by both mechanisms simultaneously. And within the central program for these movements, control simultaneously involves a small set of pacemaker neurons and larger circuits of neurons that contribute to rhythm generation through a network of synaptic connections.

If a country's government tried to simultaneously implement two competing mechanisms for decision making, there would be problems. Only one government mechanism can be allowed to make each type of decision.

But this is not the case in nervous systems. Somehow, nervous systems are able to implement distinct architectures of government simultaneously without problematic conflicts. How is this possible?

It may be that competing mechanisms can coexist in nervous systems because the goal of an animal is not to implement one particular governing structure; the goal is to produce the best available behavior—that which maximizes the probability of surviving or reproducing—at each moment. Whatever mechanisms happen to work to generate the behavior appropriately are preserved and largely replicated in the animal's offspring, even if this means there are multiple competing mechanisms acting simultaneously. A command neuron and a large population of broadly tuned neurons may simultaneously influence motor neurons. As long as the motor neurons spike at the right time and control the muscles suitably, all is well. As long as the animal is not eaten and produces many successful offspring, there is no need for a court case to determine the correct governmental mechanism.

In nervous systems, multiple coexisting mechanisms may be the rule, rather than the exception. For example, both a pacemaker neuron and a network of synaptically connected neurons appear to simultaneously control digestive movements in crustaceans.[1] Similarly, both pacemaker neurons and a network of synaptically connected neurons in the medulla may simultaneously control breathing movements in mammals.[2]

Having a single mechanism operating would certainly improve clarity of government, but nervous systems are not concerned with clarity, only with survival and reproduction. Having multiple mechanisms operating simultaneously to control each behavior provides greater reliability than using one mechanism alone. Injury to some of the key players (including a command neuron) need not incapacitate the government of behavior, even temporarily. The different mechanisms at work may not exactly be redundant—they may have somewhat different effects on the behavior—yet the animal can still get by with one mechanism in the absence of the other.

Scientists generally prefer clarity. This may help explain why there are so many disagreements in science about which mechanism is at work to control behaviors (among other topics of disagreement). Is the behavior triggered by a single command neuron or instead by a large network of broadly tuned neurons, each making a small contribution? Is a rhythmic behavior generated by a central program or instead by a sequence of sensory feedback signals, each triggering the next movement? Is a centrally generated rhythm produced by pacemaker neurons or instead by a synaptically connected network of neurons? Are neurons in central pattern generators specialized for one rhythmic behavior or instead multifunctional? Do brains keep track of an animal's own movements through corollary discharge signals or instead via sensory feedback signals from muscles and joints? Are neuronal circuits formed from genetic instructions or instead through learning?

In each of these cases, it has become clear that the correct answer is "both." But it has often required decades of lively disagreement to reach that conclusion. Perhaps a good rule of thumb would be that if we can imagine multiple mechanisms for how the nervous system controls a behavior, it probably uses all of them.

A second difference between countries' governments and nervous systems may be the capacity nervous systems have for immediate and autonomous self-correction.

Neuronal circuits are constantly changing. Some changes in the nervous system lead to an altered behavior that improves the animal's performance. In these cases, we would say that the nervous system changes are mechanisms of learning, such as we've explored for crayfish escape and birdsong.

Other changes in the nervous system actually maintain a behavior unaltered in the face of changing external or internal circumstances that influence nervous system mechanisms. In these cases, we would say the nervous system exhibits a kind of homeostasis. Behaviors, just like body temperature, blood pressure, and blood glucose and oxygen levels, can be regulated to stay within acceptable ranges. Neuronal circuits, and even individual neurons, automatically modify their mechanisms to maintain their outputs, and thereby the animal's behavior, within stable ranges that work well, as we've seen for crustacean digestion.[3] The mechanisms governing neuron and circuit homeostasis are currently being investigated by many researchers.[4]

It is hard to imagine an autonomous, self-correcting process operating with similar success and efficiency in the government of a country or the control of a company. This has been tried in computer programs that initiate lightning-fast and large-scale stock market trades. Such programs have contributed to dramatic sell-offs that generated stock market crashes.[5] We simply aren't very good at creating automatic mechanisms that take into account all the possible consequences.

But organisms are different. As we've seen, neuronal circuits often alter themselves to preserve a stable output. The scalpel of evolution has cut away mechanisms of change that often have disastrous results. Those that have survived the cut generally serve the organism's interests well.

A third difference between countries' governments and nervous systems may be the variety of ways in which different nervous systems are able to achieve the same goals approximately equally well. As we've seen, even within a particular neuronal circuit in a particular species, such as the stomatogastric system of a crab, the circuit uses different sets of mechanisms in individual animals.[6] Each set of mechanisms appears to be approximately equally successful, and one individual animal may even alter the set of mechanisms it uses during its lifetime. Thus, for nervous systems there is not one best way. Instead, diversity is a major feature of them. Your brain's best way and my brain's best way might be different, but they both may get the job done equally well.

Notes

Credits

Acknowledgments

Index

Notes

1. HOW TO SPY ON THE GOVERNMENT

1 Aristotle, *Metaphysics,* vol. 5 (Cambridge, MA: Harvard University Press, 1989).

2 N. Tinbergen, "On aims and methods of ethology," *Zeitschrift für Tierpsychologie* 20 (1963): 410–433.

3 Ibid.

4 E. A. Brenowitz and H. H. Zakon, "Emerging from the bottleneck: benefits of the comparative approach to modern neuroscience," *Trends in Neurosciences* 38 (2015): 273–278.

5 A. Krogh, "The progress of physiology," *American Journal of Physiology* 90 (1929): 243–251.

2. ISN'T THERE AN EASIER WAY?

1 L. A. Jorgenson et al., "The BRAIN Initiative: developing technology to catalyse neuroscience discovery," *Philosophical Transactions of the Royal Society of London B* 370 (2015): 20140164.

2 J. A. Egeland et al., "Bipolar affective disorders linked to DNA markers on chromosome 11," *Nature* 325 (1987): 783–787; M. Baron et al., "Genetic linkage between X-chromosome markers and bipolar affective illness," *Nature* 326 (1987): 289–292; M. Robertson, "Molecular genetics of the mind," *Nature* 325 (1987): 755; S. D. Detera-Wadleigh et al., "Close linkage of c-Harvey-ras-1 and the insulin gene to affective disorder is ruled out in three North American pedigrees," *Nature* 325 (1987): 806–808; S. Hodgkinson et al., "Molecular genetic evidence for heterogeneity in manic depression," *Nature* 325 (1987): 805–806; D. H. Hamer et al., "A linkage between DNA markers on the X chromosome and male sexual orientation," *Science* 261 (1993): 321–327; W. H. Berrettini et al., "Chromosome 18 DNA markers and manic-depressive illness: Evidence for a susceptibility gene," *Proceedings of the National Academy of Science USA* 91 (1994): 5918–5921; R. E. Straub et al., "A possible vulnerability locus for bipolar affective disorder on chromosome 21q22.3," *Nature Genetics* 8 (1994): 291–296.

3 J. L. Kennedy et al., "Evidence against linkage of schizophrenia to markers on chromosome 5 in a northern Swedish pedigree," *Nature* 336 (1988): 167–170; S. Peele, "Second thoughts about a gene for alcoholism," *The Atlantic Monthly* 266 (1990): 52–58; J. Gelernter et al., "No association between an allele at the D2 dopamine receptor gene (DRD2) and alcoholism," *Journal of the American Medical Association* 266 (1991): 1801–1807; J. S. Alper and M. R. Natowicz, "On establishing the genetic basis of mental disease," *Trends in Neurosciences* 16 (1993): 387–389; M. Baron et al., "Diminished support for linkage between manic depressive illness and X-chromosome markers in three Israeli pedigrees," *Nature Genetics* 3 (1993): 49–55; J. Gelernter et al., "The A1 allele at the D2 dopamine receptor gene and alcoholism: A reappraisal," *Journal of the American Medical Association* 269 (1993): 1673–1677; C. R. Cloninger, "Turning point in the design of linkage studies of schizophrenia," *American Journal of Medical Genetics* 54 (1994): 83–92; E. S. Gershon and C. R. Cloninger, *Genetic Approaches to Mental Disorders* (Washington, DC: American Psychiatric Press, 1994); A. Berkowitz, "Our genes, ourselves?," *BioScience* 46 (1996): 42–51; D. Hamer, "Genetics: Rethinking behavior genetics," *Science* 298 (2002): 71–72; B. S. Mustanski et al., "A genomewide scan of male sexual orientation," *Human Genetics* 116 (2005): 272–278; S. A. McCarroll et al., "Genome-scale neurogenetics: Methodology and meaning," *Nature Neuroscience* 17 (2014): 756–763; A. C. Need and D. B. Goldstein, "Schizophrenia genetics comes of age," *Neuron* 83 (2014): 760–763.
4 Hamer, "Genetics"; P. A. Arguello and J. A. Gogos, "Genetic and cognitive windows into circuit mechanisms of psychiatric disease," *Trends in Neurosciences* 35 (2012): 3–13; McCarroll et al., "Genome-scale neurogenetics."
5 K. J. Mitchell, "The miswired brain: Making connections from neurodevelopment to psychopathology," *BMC Biology* 9 (2011): 23.
6 Hamer, "Genetics"; T. A. Manolio et al., "Finding the missing heritability of complex diseases," *Nature* 461 (2009): 747–753; Arguello and Gogos, "Genetic and cognitive windows"; J. Kaiser, "Human genetics: Genetic influences on disease remain hidden," *Science* 338 (2012): 1016–1017; J. Flint and M. R. Munafo, "Candidate and non-candidate genes in behavior genetics," *Current Opinion in Neurobiology* 23 (2013): 57–61; McCarroll et al., "Genome-scale neurogenetics"; Need and Goldstein, "Schizophrenia genetics comes of age."
7 J. Flint and M. R. Munafo, "Genetics: Finding genes for schizophrenia," *Current Biology* 24 (2014): R755–R757; Need and Goldstein, "Schizophrenia genetics comes of age"; E. L. Heinzen et al., "The

genetics of neuropsychiatric diseases: Looking in and beyond the exome," *Annual Review of Neuroscience* 38 (2015): 47–68.

8 Arguello and Gogos, "Genetic and cognitive windows."

9 K. K. Siwicki and E. A. Kravitz, "*Fruitless, doublesex* and the genetics of social behavior in *Drosophila melanogaster*," *Current Opinion in Neurobiology* 19 (2009): 200–206; A. C. von Philipsborn et al., "Neuronal control of *Drosophila* courtship song," *Neuron* 69 (2011): 509–522; J. Kohl et al., "A bidirectional circuit switch reroutes pheromone signals in male and female brains," *Cell* 155 (2013): 1610–1623.

10 L. A. McGraw and L. J. Young, "The prairie vole: An emerging model organism for understanding the social brain," *Trends in Neurosciences* 33 (2010): 103–109.

11 H. E. Ross and L. J. Young, "Oxytocin and the neural mechanisms regulating social cognition and affiliative behavior," *Frontiers in Neuroendocrinology* 30 (2009): 534–547; D. T. Blumstein et al., "Toward an integrative understanding of social behavior: New models and new opportunities," *Frontiers in Behavioral Neuroscience* 4 (2010): 34; McGraw and Young, "The prairie vole"; A. Meyer-Lindenberg et al., "Oxytocin and vasopressin in the human brain: Social neuropeptides for translational medicine," *Nature Reviews Neuroscience* 12 (2011): 524–538; L. J. Young and B. Alexander, *The Chemistry between Us: Love, Sex, and the Science of Attraction* (London: Penguin, 2012); J. D. Klatt and J. L. Goodson, "Oxytocin-like receptors mediate pair bonding in a socially monogamous songbird," *Proceedings of the Royal Society of London—Series B: Biological Sciences* 280 (2013): 20122396.

12 On Huntington's disease, see I. Shoulson and A. B. Young, "Milestones in Huntington disease," *Movement Disorders* 26 (2011): 1127–1133; on Parkinson's disease, see T. Gasser et al., "Milestones in PD genetics," *Movement Disorders* 26 (2011): 1042–1048; on Alzheimer's disease, see K. Bettens et al., "Current status on Alzheimer disease molecular genetics: From past, to present, to future," *Human Molecular Genetics* 19 (2010): R4–R11; R. E. Tanzi, "The genetics of Alzheimer disease," *Cold Spring Harbor Perspectives in Medicine* 2 (2012), doi: 10.1101/cshperspect.a006296.

13 D. J. Heeger and D. Ress, "What does fMRI tell us about neuronal activity?," *Nature Reviews Neuroscience* 3 (2002): 142–151; S. G. Kim and S. Ogawa, "Insights into new techniques for high resolution functional MRI," *Current Opinion in Neurobiology* 12 (2002): 607–615; N. K. Logothetis, "The neural basis of the blood-oxygen-level-dependent functional magnetic resonance imaging signal," *Philosophical Transactions of the Royal Society of London B: Biological*

Sciences 357 (2002): 1003–1037; R. C. deCharms, "Applications of real-time fMRI," *Nature Reviews Neuroscience* 9 (2008): 720–729; H. Shibasaki, "Human brain mapping: Hemodynamic response and electrophysiology," *Clinical Neurophysiology* 119 (2008): 731–743; D. J. McKeefry et al., "The noninvasive dissection of the human visual cortex: Using fMRI and TMS to study the organization of the visual brain," *Neuroscientist* 15 (2009): 489–506; H. U. Voss and N. D. Schiff, "MRI of neuronal network structure, function, and plasticity," *Progress in Brain Research* 175 (2009): 483–496; C. M. Bennett and M. B. Miller, "How reliable are the results from functional magnetic resonance imaging?," *Annals of the New York Academy of Sciences* 1191 (2010): 133–155; A. Ekstrom, "How and when the fMRI BOLD signal relates to underlying neural activity: The danger in dissociation," *Brain Research Reviews* 62 (2010): 233–244; P. A. Bandettini, "Twenty years of functional MRI: The science and the stories," *Neuroimage* 62 (2012): 575–588; E. M. Hillman, "Coupling mechanism and significance of the BOLD signal: A status report," *Annual Review of Neuroscience* 37 (2014): 161–181; J. Dubois et al., "Single-unit recordings in the macaque face patch system reveal limitations of fMRI MVPA," *Journal of Neuroscience* 35 (2015): 2791–2802; P. J. Magistretti and I. Allaman, "A cellular perspective on brain energy metabolism and functional imaging," *Neuron* 86 (2015): 883–901.

14 E. Zarahn, "Spatial localization and resolution of BOLD fMRI," *Current Opinion in Neurobiology* 11 (2001): 209–212; Kim and Ogawa, "Insights into new techniques"; Logothetis, "The neural basis"; Shibasaki, "Human brain mapping"; J. Goense et al., "High-resolution fMRI reveals laminar differences in neurovascular coupling between positive and negative BOLD responses," *Neuron* 76 (2012): 629–639.

15 L. V. Wang, "Photo acoustic tomography," *Scholarpedia* 9 (2014): 10278, doi: 10.4249/scholarpedia.10278.

16 S. Seung, *Connectome: How the Brain's Wiring Makes Us Who We Are* (New York: Houghton Mifflin Harcourt, 2012).

17 S. Faumont et al., "Neuronal microcircuits for decision making in C. elegans," *Current Opinion in Neurobiology* 22 (2012): 580–591.

18 S. Finger, *Origins of Neuroscience* (New York: Oxford University Press, 1994).

19 J.-B. Bouillaud, "Recherches cliniques propres à démontrer que la perte de la parole correspond à la lésion des lobules antérieurs du cerveau, et à confirmer l'opinion de Gall, sur le siège du langage articulé," *Archives Génerale de Médecine (Paris)* 8 (1825): 25–45; P. Broca, "Remarques sur le siège de la faculté du langage articulé; suivies

d'une observation d'aphémie (perte de la parole)" [Remarks on the seat of the faculty of articulate language, followed by an observation of aphemia], *Bulletins de la Société Anatomique de Paris* 6 (1861): 330–357, in *Some Papers on the Cerebral Cortex*, ed. and trans. Gerhardt von Bonin (Springfield, IL: Charles C. Thomas, 1960), 49–72; Finger, *Origins of Neuroscience*.

20 Finger, *Origins of Neuroscience*.

21 Ibid.

22 D. P. McCabe and A. D. Castel, "Seeing is believing: The effect of brain images on judgments of scientific reasoning," *Cognition* 107 (2008): 343–352.

23 K. S. Lashley, "Basic neural mechanisms in behavior," *Psychological Review* 37 (1930): 1–24.

24 Ibid.

3. NEURONAL DICTATORSHIPS

1 I. Kupfermann and K. R. Weiss, "The command neuron concept," *The Behavioral and Brain Sciences* 1 (1978): 3–39.

2 C. A. G. Wiersma and K. Ikeda, "Interneurons commanding swimmeret movements in the crayfish, *Procambarus Clarki* (Girard)," *Comparative Biochemistry and Physiology* 12 (1964): 509–525.

3 D. H. Edwards et al., "Fifty years of a command neuron: The neurobiology of escape behavior in the crayfish," *Trends in Neurosciences* 22 (1999): 153–161; R. C. Eaton et al., "The Mauthner cell and other identified neurons of the brainstem escape network of fish," *Progress in Neurobiology* 63 (2001): 467–485.

4 S. J. Zottoli, "Correlation of the startle reflex and Mauthner cell auditory responses in unrestrained goldfish," *Journal of Experimental Biology* 66 (1977): 243–254; R. C. Eaton et al., "Identification of Mauthner-initiated response patterns in goldfish: Evidence from simultaneous cinematography and electrophysiology," *Journal of Comparative Physiology* 144 (1981): 521–531.

5 M. K. Rock, "Functional properties of Mauthner cell in the tadpole *Rana catesbeiana*," *Journal of Neurophysiology* 44 (1980): 135–150; M. K. Rock et al., "Does the Mauthner cell conform to the criteria of the command neuron concept?," *Brain Research* 204 (1981): 21–27; J. Nissanov et al., "The motor output of the Mauthner cell, a reticulospinal command neuron," *Brain Research* 517 (1990): 88–98.

6 Rock et al., "Does the Mauthner cell conform?"

7 R. C. Eaton et al., "Alternative neural pathways initiate fast-start responses following lesions of the Mauthner neuron in goldfish," *Journal of Comparative Physiology* 145 (1982): 485–496.

8 R. DiDomenico et al., "Lateralization and adaptation of a continuously variable behavior following lesions of a reticulospinal command neuron," *Brain Research* 473 (1988): 15–28.

9 Nissanov et al., "Motor output of the Mauthner cell"; R. C. Eaton et al., "Role of the Mauthner cell in sensorimotor integration by the brain stem escape network," *Brain, Behavior and Evolution* 37 (1991): 272–285; D. M. O'Malley et al., "Imaging the functional organization of zebrafish hindbrain segments during escape behaviors," *Neuron* 17 (1996): 1145–1155; J. L. Casagrand et al., "Mauthner and reticulospinal responses to the onset of acoustic pressure and acceleration stimuli," *Journal of Neurophysiology* 82 (1999): 1422–1437; K. S. Liu and J. R. Fetcho, "Laser ablations reveal functional relationships of segmental hindbrain neurons in zebrafish," *Neuron* 23 (1999): 325–335; S. J. Zottoli et al., "Decrease in occurrence of fast startle responses after selective Mauthner cell ablation in goldfish *(Carassius auratus),*" *Journal of Comparative Physiology A—Sensory, Neural, and Behavioral Physiology* 184 (1999): 207–218; E. Gahtan et al., "Evidence for a widespread brain stem escape network in larval zebrafish," *Journal of Neurophysiology* 87 (2002): 608–614.

10 C. A. Wiersma, "Giant nerve fiber system of the crayfish: A contribution to comparative physiology of synapse," *Journal of Neurophysiology* 10 (1947): 23–38; J. J. Wine and F. B. Krasne, "The organization of escape behaviour in the crayfish," *Journal of Experimental Biology* 56 (1972): 1–18; Edwards et al., "Fifty years of a command neuron."

11 J. E. Schrameck, "Crayfish swimming: Alternating motor output and giant fiber activity," *Science* 169 (1970): 698–700; Wine and Krasne, "Organization of escape behaviour"; A. P. Kramer et al., "Different command neurons select different outputs from a shared premotor interneuron of crayfish tail-flip circuitry," *Science* 214 (1981): 810–812; A. P. Kramer and F. B. Krasne, "Crayfish escape behavior: Production of tailflips without giant fiber activity," *Journal of Neurophysiology* 52 (1984): 189–211.

12 Wine and Krasne, "Organization of escape behaviour"; Edwards et al., "Fifty years of a command neuron."

13 G. C. Olson and F. B. Krasne, "The crayfish lateral giants as command neurons for escape behavior," *Brain Research* 214 (1981): 89–100.

14 A. Roberts et al., "Segmental giant: Evidence for a driver neuron interposed between command and motor neurons in the crayfish escape system," *Journal of Neurophysiology* 47 (1982): 761–781.

15 A. Roberts, "Recurrent inhibition in the giant-fibre system of the crayfish and its effect on the excitability of the escape response," *Journal of Experimental Biology* 48 (1968): 545–567.

16 F. B. Krasne, "Excitation and habituation of the crayfish escape reflex: The depolarizing response in lateral giant fibres of the isolated abdomen," *Journal of Experimental Biology* 50 (1969): 29–46; F. B. Krasne and K. S. Woodsmall, "Waning of the crayfish escape response as a result of repeated stimulation," *Animal Behaviour* 17 (1969): 416–424; J. J. Wine et al., "Habituation and inhibition of the crayfish lateral giant fibre escape response," *Journal of Experimental Biology* 62 (1975): 771–782.

17 Krasne, "Excitation and habituation of the crayfish"; Krasne and Woodsmall, "Waning of the crayfish escape response"; Wine et al., "Habituation and inhibition of the crayfish."

18 F. B. Krasne and J. S. Bryan, "Habituation: Regulation through presynaptic inhibition," *Science* 182 (1973): 590–592.

19 E. T. Vu et al., "Postexcitatory inhibition of the crayfish lateral giant neuron: A mechanism for sensory temporal filtering," *Journal of Neuroscience* 17 (1997): 8867–8879; J. S. Bryan and F. B. Krasne, "Protection from habituation of the crayfish lateral giant fibre escape response," *Journal of Physiology* 271 (1977): 351–368; J. S. Bryan and F. B. Krasne, "Presynaptic inhibition: The mechanism of protection from habituation of the crayfish lateral giant fibre escape response," *Journal of Physiology* 271 (1977): 369–390; M. D. Kirk, "Presynaptic inhibition in the crayfish CNS: Pathways and synaptic mechanisms," *Journal of Neurophysiology* 54 (1985): 1305–1325.

20 S. R. Yeh et al., "The effect of social experience on serotonergic modulation of the escape circuit of crayfish," *Science* 271 (1996): 366–369; F. B. Krasne et al., "Altered excitability of the crayfish lateral giant escape reflex during agonistic encounters," *Journal of Neuroscience* 17 (1997): 709–716.

21 Krasne et al., "Altered excitability of the crayfish."

22 Yeh et al., "Effect of social experience"; Krasne et al., "Altered excitability of the crayfish."

23 Yeh et al., "Effect of social experience."

24 On mollusks, see W. N. Frost and P. S. Katz, "Single neuron control over a complex motor program," *Proceedings of the National Academy of Science USA* 93 (1996): 422–426; on leeches, see J. G. Puhl et al., "Necessary, sufficient and permissive: A single locomotor command neuron important for intersegmental coordination," *Journal of Neuroscience* 32 (2012): 17646–17657.

25 C. R. von Reyn et al., "A spike-timing mechanism for action selection," *Nature Neuroscience* 17 (2014): 962–970.

26 On feeding in fruit flies, see T. F. Flood et al., "A single pair of interneurons commands the *Drosophila* feeding motor program," *Nature*

499 (2013): 83–87; on stridulation in crickets, see B. Hedwig, "Control of cricket stridulation by a command neuron: Efficacy depends on the behavioral state," *Journal of Neurophysiology* 83 (2000): 712–722.

27 W. James, *The Principles of Psychology* (Cambridge, MA: Harvard University Press, 1983).

28 C. S. Sherrington, *Man on His Nature* (New York: The Macmillan Company, 1941).

29 H. B. Barlow, "Summation and inhibition in the frog's retina," *Journal of Physiology* 119 (1953): 69–88.

30 J. Y. Lettvin et al., "What the frog's eye tells the frog's brain," *Proceedings of the Institute of Radio Engineers* 47 (1959): 1940–1951.

31 H. B. Barlow, "Single units and sensation: A neuron doctrine for perceptual psychology?," *Perception* 1 (1972): 371–394.

32 C. G. Gross, "Genealogy of the 'grandmother cell,'" *Neuroscientist* 8 (2002): 512–518.

33 H. Barlow, "The neuron doctrine in perception," in *The Cognitive Neurosciences*, ed. Michael S. Gazzaniga, 415–435 (Cambridge, MA: MIT Press, 1994).

34 C. G. Gross et al., "Visual properties of neurons in inferotemporal cortex of the macaque," *Journal of Neurophysiology* 35 (1972): 96–111; D. I. Perrett et al., "Visual neurones responsive to faces in the monkey temporal cortex," *Experimental Brain Research* 47 (1982): 329–342; R. Desimone et al., "Stimulus-selective properties of inferior temporal neurons in the macaque," *Journal of Neuroscience* 4 (1984): 2051–2062; C. G. Gross and J. Sergent, "Face recognition," *Current Opinion in Neurobiology* 2 (1992): 156–161; M. P. Young and S. Yamane, "Sparse population coding of faces in the inferotemporal cortex," *Science* 256 (1992): 1327–1331.

35 W. Penfield and E. Boldrey, "Somatic motor and sensory representation in the cerebral cortex of man as studied by electrical stimulation," *Brain* 60 (1937): 389–443.

36 R. Q. Quiroga et al., "Invariant visual representation by single neurons in the human brain," *Nature* 435 (2005): 1102–1107; R. Q. Quiroga et al., "Human single-neuron responses at the threshold of conscious recognition," *Proceedings of the National Academy of Science USA* 105 (2008): 3599–3604; R. Q. Quiroga et al., "Explicit encoding of multimodal percepts by single neurons in the human brain," *Current Biology* 19 (2009): 1308–1313; R. Q. Quiroga, "Concept cells: The building blocks of declarative memory functions," *Nature Reviews Neuroscience* 13 (2012): 587–597.

37 Quiroga et al., "Explicit encoding of multimodal percepts."

38 W. T. Newsome, "The King Solomon Lectures in Neuroethology: Deciding about motion: Linking perception to action," *Journal of Comparative Physiology A—Sensory, Neural, and Behavioral Physiology* 181 (1997): 5–12; A. J. Parker and W. T. Newsome, "Sense and the single neuron: Probing the physiology of perception," *Annual Review of Neuroscience* 21 (1998): 227–277.

39 K. H. Britten et al., "The analysis of visual motion: A comparison of neuronal and psychophysical performance," *Journal of Neuroscience* 12 (1992): 4745–4765.

40 S. Celebrini and W. T. Newsome, "Neuronal and psychophysical sensitivity to motion signals in extrastriate area MST of the macaque monkey," *Journal of Neuroscience* 14 (1994): 4109–4124.

41 C. D. Salzman and W. T. Newsome, "Neural mechanisms for forming a perceptual decision," *Science* 264 (1994): 231–237.

4. NEURONAL DEMOCRACIES

1 T. Young, "The Bakerian lecture: On the theory of light and colours," *Philosophical Transactions of the Royal Society of London* 92 (1802): 12–48.

2 Ibid.

3 E. D. Adrian, *The Basis of Sensation* (New York: W. W. Norton and Company, 1928).

4 G. E. Hinton et al., "Distributed representations," in *Parallel Distributed Processing: Exploration in the Microstructure of Cognition*, vol. 1: *Foundations*, ed. D. E. Rumelhart and J. L. McClelland (Cambridge, MA: MIT Press, 1986), 77–109; W. Heiligenberg, "Central processing of sensory information in electric fish," *Journal of Comparative Physiology A—Sensory, Neural, and Behavioral Physiology* 161 (1987): 621–631; P. Baldi and W. Heiligenberg, "How sensory maps could enhance resolution through ordered arrangements of broadly tuned receivers," *Biological Cybernetics* 59 (1988): 313–318; F. E. Theunissen and J. P. Miller, "Representation of sensory information in the cricket cercal sensory system: II. Information theoretic calculation of system accuracy and optimal tuning-curve widths of four primary interneurons," *Journal of Neurophysiology* 66 (1991): 1690–1703; S. Deneve et al., "Reading population codes: A neural implementation of ideal observers," *Nature Neuroscience* 2 (1999): 740–745; T. D. Sanger, "Neural population codes," *Current Opinion in Neurobiology* 13 (2003): 238–249.

5 W. Heiligenberg, "Jamming avoidance responses: Model systems for neuroethology," in *Electroreception*, ed. T. H. Bullock and W. Heiligenberg (New York: John Wiley and Sons, 1986), 613–649; G. J. Rose,

"Insights into neural mechanisms and evolution of behaviour from electric fish," *Nature Reviews Neuroscience* 5 (2004): 943–951.

6 Heiligenberg, "Jamming avoidance responses."

7 J. P. Miller et al., "Representation of sensory information in the cricket cercal sensory system: I. Response properties of the primary interneurons," *Journal of Neurophysiology* 66 (1991): 1680–1689.

8 Theunissen and Miller, "Representation of sensory information: II."

9 J. E. Lewis and W. B. Kristan Jr., "A neuronal network for computing population vectors in the leech," *Nature* 391 (1998): 76–79.

10 D. L. Sparks et al., "Size and distribution of movement fields in the monkey superior colliculus," *Brain Research* 113 (1976): 21–34.

11 D. A. Robinson, "Eye movements evoked by collicular stimulation in the alert monkey," *Vision Research* 12 (1972): 1795–1808.

12 C. Lee et al., "Population coding of saccadic eye movements by neurons in the superior colliculus," *Nature* 332 (1988): 357–360.

13 A. P. Georgopoulos et al., "On the relations between the direction of two-dimensional arm movements and cell discharge in primate motor cortex," *Journal of Neuroscience* 2 (1982): 1527–1537.

14 Ibid.; A. P. Georgopoulos et al., "Neuronal population coding of movement direction," *Science* 233 (1986): 1416–1419.

15 On escape direction in cockroaches, see E. Liebenthal et al., "Critical parameters of the spike trains in a cell assembly: Coding of turn direction by the giant interneurons of the cockroach," *Journal of Comparative Physiology A—Sensory, Neural, and Behavioral Physiology* 174 (1994): 281–296; R. Levi and J. M. Camhi, "Wind direction coding in the cockroach escape response: Winner does not take all," *Journal of Neuroscience* 20 (2000): 3814–3821; on odor identity in insects, see R. W. Friedrich and M. Stopfer, "Recent dynamics in olfactory population coding," *Current Opinion in Neurobiology* 11 (2001): 468–474; R. A. Campbell et al., "Imaging a population code for odor identity in the *Drosophila* mushroom body," *Journal of Neuroscience* 33 (2013): 10568–10581; on a fly's motion while flying, see K. Karmeier et al., "Population coding of self-motion: Applying Bayesian analysis to a population of visual interneurons in the fly," *Journal of Neurophysiology* 94 (2005): 2182–2194; on sound frequency in fish, see D. A. Bodnar et al., "Temporal population code of concurrent vocal signals in the auditory midbrain," *Journal of Comparative Physiology A—Sensory, Neural, and Behavioral Physiology* 187 (2001): 865–873; on body location touched in turtles, see A. Berkowitz and P. S. Stein, "Activity of descending propriospinal axons in the turtle hindlimb enlargement during two forms of fictive scratching: Broad tuning to regions of the body surface," *Journal of Neuroscience* 14 (1994):

5089–5104; A. Berkowitz, "Broadly tuned spinal neurons for each form of fictive scratching in spinal turtles," *Journal of Neurophysiology* 86 (2001): 1017–1025; on sound source location in owls, see M. Konishi, "Coding of auditory space," *Annual Review of Neuroscience* 26 (2003): 31–55; on sound parameters in bats, see E. Covey, "Neural population coding and auditory temporal pattern analysis," *Physiology & Behavior* 69 (2000): 211–220; on whiskers in rats, see R. S. Petersen et al., "Population coding of stimulus location in rat somatosensory cortex," *Neuron* 32 (2001): 503–514; on tastes in mammals, see T. R. Scott and B. K. Giza, "Issues of gustatory neural coding: Where they stand today," *Physiology & Behavior* 69 (2000): 65–76; D. V. Smith et al., "Neuronal cell types and taste quality coding," *Physiology & Behavior* 69 (2000): 77–85; on somatosensory qualities in mammals, see G. S. Doetsch, "Patterns in the brain: Neuronal population coding in the somatosensory system," *Physiology & Behavior* 69 (2000): 187–201; on object perception in ferrets, see A. Basole et al., "Mapping multiple features in the population response of visual cortex," *Nature* 423 (2003): 986–990; on leg position in cats, see G. Bosco and R. E. Poppele, "Broad directional tuning in spinal projections to the cerebellum," *Journal of Neurophysiology* 70 (1993): 863–866; on arm movements in monkeys, see W. T. Thach, "Correlation of neural discharge with pattern and force of muscular activity, joint position, and direction of intended next movement in motor cortex and cerebellum," *Journal of Neurophysiology* 41 (1978): 654–676; M. T. Johnson et al., "Central processes for the multiparametric control of arm movements in primates," *Current Opinion in Neurobiology* 11 (2001): 684–688; on the lengths of muscles in humans, see M. Bergenheim et al., "Proprioceptive population coding of two-dimensional limb movements in humans: I. Muscle spindle feedback during spatially oriented movements," *Experimental Brain Research* 134 (2000): 301–310.

16 M. A. Nicolelis and M. A. Lebedev, "Principles of neural ensemble physiology underlying the operation of brain-machine interfaces," *Nature Reviews Neuroscience* 10 (2009): 530–540; A. M. Green and J. F. Kalaska, "Learning to move machines with the mind," *Trends in Neurosciences* 34 (2011): 61–75; N. G. Hatsopoulos and A. J. Suminski, "Sensing with the motor cortex," *Neuron* 72 (2011): 477–487; M. A. Nicolelis, *Beyond Boundaries* (New York: Times Books, 2011).

17 A. P. Georgopoulos et al., "Primate motor cortex and free arm movements to visual targets in three-dimensional space: II. Coding of the direction of movement by a neuronal population," *Journal of Neuroscience* 8 (1988): 2928–2937.

18 L. R. Hochberg et al., "Reach and grasp by people with tetraplegia using a neurally controlled robotic arm," *Nature* 485 (2012): 372–375.

19 C. T. Moritz et al., "Direct control of paralysed muscles by cortical neurons," *Nature* 456 (2008): 639–642.

20 T. Aflalo et al., "Decoding motor imagery from the posterior parietal cortex of a tetraplegic human," *Science* 348 (2015): 906–910.

5. HOW ARE THE FACTORIES RUN?

1 H. F. Brown et al., "How does adrenaline accelerate the heart?," *Nature* 280 (1979): 235–236; M. Biel et al., "Hyperpolarization-activated cation channels: From genes to function," *Physiological Reviews* 89 (2009): 847–885; D. DiFrancesco, "The role of the funny current in pacemaker activity," *Circulation Research* 106 (2010): 434–446.

2 O. Monfredi et al., "Modern concepts concerning the origin of the heartbeat," *Physiology (Bethesda)* 28 (2013): 74–92.

3 W. B. Cannon, "The story of the development of our ideas of chemical mediation of nerve impulses," *The American Journal of the Medical Sciences* 188 (1934): 145–159.

4 C. S. Sherrington, *The Integrative Action of the Nervous System* (New York: Charles Scribner's Sons, 1906); C. S. Sherrington, "Flexion-reflex of the limb, crossed extension-reflex, and reflex stepping and standing," *Journal of Physiology* 40 (1910): 28–121.

5 C. S. Sherrington, "Observations on the scratch-reflex in the spinal dog," *Journal of Physiology* 34 (1906): 1–50; C. S. Sherrington, "Notes on the scratch-reflex of the cat," *Quarterly Journal of Experimental Physiology* 3 (1910): 213–220.

6 T. G. Brown, "The intrinsic factors in the act of progression in the mammal," *Proceedings of the Royal Society of London* 84 (1911): 308–319.

7 Ibid.

8 Ibid.

9 E. Marder and R. L. Calabrese, "Principles of rhythmic motor pattern generation," *Physiological Reviews* 76 (1996): 687–717; P. S. G. Stein et al., *Neurons, Networks, and Motor Behavior* (Cambridge, MA: MIT Press, 1997); O. Kiehn et al., *Neuronal Mechanisms for Generating Locomotor Activity* (New York: New York Academy of Sciences, 1998).

10 On spontaneous walking movements, see G. Holmes, "The Goulstonian Lectures on Spinal Injuries of Warfare: Delivered before the Royal College of Physicians of London," *British Medical Journal* 2 (1915): 815–821; B. Bussel et al., "Myoclonus in a patient with spinal cord transection: Possible involvement of the spinal stepping gener-

ator," *Brain* 111 (Pt 5) (1988): 1235–1245; B. Calancie et al., "Involuntary stepping after chronic spinal cord injury: Evidence for a central rhythm generator for locomotion in man," *Brain* 117 (Pt 5) (1994): 1143–1159; on walking movements performed on treadmills, see V. Dietz et al., "Locomotor activity in spinal man," *Lancet* 344 (1994): 1260–1263; on walking movements caused by electrical stimulation of nerves or the spinal cord, see K. Minassian et al., "Stepping-like movements in humans with complete spinal cord injury induced by epidural stimulation of the lumbar cord: Electromyographic study of compound muscle action potentials," *Spinal Cord* 42 (2004): 401–416; K. Minassian et al., "Human lumbar cord circuitries can be activated by extrinsic tonic input to generate locomotor-like activity," *Human Movement Science* 26 (2007): 275–295; on walking movements performed while sleeping, see T. Yokota et al., "Sleep-related periodic leg movements (nocturnal myoclonus) due to spinal cord lesion," *Journal of the Neurological Sciences* 104 (1991): 13–18.

11 Sherrington, "Flexion-reflex of the limb."

12 D. G. Lamb and R. L. Calabrese, "Neural circuits controlling behavior and autonomic functions in medicinal leeches," *Neural Systems and Circuits* 1 (2011): 13.

13 On walking in mammals, see M. Hagglund et al., "Optogenetic dissection reveals multiple rhythmogenic modules underlying locomotion," *Proceedings of the National Academy of Science USA* 110 (2013): 11589–11594; on other rhythmic behavoirs, see J. A. Kahn and A. Roberts, "Experiments on the central pattern generator for swimming in amphibian embryos," *Philosophical Transactions of the Royal Society of London, Series B: Biological Sciences* 296 (1982): 229–243; L. Cangiano and S. Grillner, "Fast and slow locomotor burst generation in the hemispinal cord of the lamprey," *Journal of Neurophysiology* 89 (2003): 2931–2942; L. Cangiano and S. Grillner, "Mechanisms of rhythm generation in a spinal locomotor network deprived of crossed connections: The lamprey hemicord," *Journal of Neuroscience* 25 (2005): 923–935; P. S. Stein, "Motor pattern deletions and modular organization of turtle spinal cord," *Brain Research Reviews* 57 (2008): 118–124; W. C. Li et al., "Specific brainstem neurons switch each other into pacemaker mode to drive movement by activating NMDA receptors," *Journal of Neuroscience* 30 (2010): 16609–16620; W. C. Li, "Generation of locomotion rhythms without inhibition in vertebrates: The search for pacemaker neurons," *Integrative and Comparative Biology* 51 (2011): 879–889; L. Cangiano et al., "The hemisegmental locomotor network revisited," *Neuroscience* 210 (2012): 33–37.

14 J. G. Jones et al., "Thomas Graham Brown (1882–1965): Behind the scenes at the Cardiff Institute of Physiology," *Journal of the History of the Neurosciences* 20 (2011): 188–209.

15 T. Weis-Fogh, "Biology and physics of locust flight: IV. Notes on sensory mechanisms in locust flight," *Philosophical Transactions of the Royal Society of London, Series B: Biological Sciences* 239 (1956): 553–584.

16 D. M. Wilson, "The central nervous control of flight in a locust," *Journal of Experimental Biology* 38 (1961): 471–490.

17 Ibid.

18 Marder and Calabrese, "Principles of rhythmic motor pattern generation"; F. Delcomyn, "Neural basis of rhythmic behavior in animals," *Science* 210 (1980): 492–498.

19 A. Lundberg, "Half-centres revisited," in *Advanced Physiological Science,* vol. 1, *Regulatory Functions of the CNS: Motion and Organization Principles,* ed. J. Szentagotheu, M. Palkovits, and J. Hamori (Budapest, Hungary: Pergamon Akademiai Kiado, 1981), 155–167; D. G. Stuart and H. Hultborn, "Thomas Graham Brown (1882–1965), Anders Lundberg (1920–), and the neural control of stepping," *Brain Research Reviews* 59 (2008): 74–95.

20 S. N. Currie, "Donald M. Wilson: The point that must be reached (1932–1970)," http://faculty.ucr.edu/~currie/donald-wilson.htm; S. N. Currie, "Donald M. Wilson (1932–1970): 'The point that must be reached,'" *International Society for Neuroethology Newsletter,* March 2011, 9–13.

21 Currie, "Donald M. Wilson: The point that must be reached (1932–1970)"; Currie, "Donald M. Wilson (1932–1970): 'The point that must be reached.'"

22 S. Finger, *Origins of Neuroscience* (New York: Oxford University Press, 1994).

23 J. C. Smith and J. L. Feldman, "In vitro brainstem–spinal cord preparations for study of motor systems for mammalian respiration and locomotion," *Journal of Neuroscience Methods* 21 (1987): 321–333.

24 J. C. Smith et al., "Pre-Bötzinger complex: A brainstem region that may generate respiratory rhythm in mammals," *Science* 254 (1991): 726–729.

25 W. F. Boron and E. L. Boulpaep, *Medical Physiology,* 2nd ed., international ed.. (Philadelphia: Saunders, 2009).

26 J. C. Smith et al., "Structural and functional architecture of respiratory networks in the mammalian brainstem," *Philosophical Transactions of the Royal Society of London B: Biological Sciences* 364 (2009): 2577–2587; A. J. Garcia III et al., "Networks within networks: The

neuronal control of breathing," *Progress in Brain Research* 188 (2011): 31–50; J. L. Feldman et al., "Understanding the rhythm of breathing: So near, yet so far," *Annual Review of Physiology* 75 (2013): 423–452; J. L. Feldman and K. Kam, "Facing the challenge of mammalian neural microcircuits: Taking a few breaths may help," *Journal of Physiology* 593 (2014): 3–23.

27 Smith et al., "Pre-Bötzinger complex"; M. Thoby-Brisson and J. M. Ramirez, "Identification of two types of inspiratory pacemaker neurons in the isolated respiratory neural network of mice," *Journal of Neurophysiology* 86 (2001): 104–112; F. Pena et al., "Differential contribution of pacemaker properties to the generation of respiratory rhythms during normoxia and hypoxia," *Neuron* 43 (2004): 105–117.

28 J. F. Paton and D. W. Richter, "Role of fast inhibitory synaptic mechanisms in respiratory rhythm generation in the maturing mouse," *Journal of Physiology* 484 (Pt 2) (1995): 505–521.

29 Smith et al., "Pre-Bötzinger complex"; Thoby-Brisson and Ramirez, "Two types of inspiratory pacemaker neurons"; Pena et al., "Differential contribution of pacemaker properties."

30 C. A. Del Negro et al., "Respiratory rhythm: An emergent network property?," *Neuron* 34 (2002): 821–830; H. Koizumi and J. C. Smith, "Persistent Na^+ and K^+-dominated leak currents contribute to respiratory rhythm generation in the pre-Bötzinger complex in vitro," *Journal of Neuroscience* 28 (2008): 1773–1785.

31 W. A. Janczewski et al., "Role of inhibition in respiratory pattern generation," *Journal of Neuroscience* 33 (2013): 5454–5465; see also D. Sherman et al., "Optogenetic perturbation of preBötzinger complex inhibitory neurons modulates respiratory pattern," *Nature Neuroscience* 18 (2015) 408–414.

32 W. A. Janczewski and J. L. Feldman, "Distinct rhythm generators for inspiration and expiration in the juvenile rat," *Journal of Physiology* 570 (2006): 407–420.

33 Feldman et al., "Understanding the rhythm of breathing."

6. THE PLOT (AND THE CHEMICAL SOUP) THICKENS

1 D. M. Maynard, "Simpler networks," *Annals of the New York Academy of Sciences* 193 (1972): 59–72.

2 R. M. Harris-Warrick et al., *Dynamic Biological Networks: The Stomatogastric Nervous System* (Cambridge, MA: The MIT Press, 1992); E. Marder and D. Bucher, "Understanding circuit dynamics using the stomatogastric nervous system of lobsters and crabs," *Annual Review of Physiology* 69 (2007): 291–316.

3 Harris-Warrick et al., *Dynamic Biological Networks;* Marder and Bucher, "Understanding circuit dynamics"; W. Stein, "Modulation of stomatogastric rhythms," *Journal of Comparative Physiology A— Sensory, Neural, and Behavioral Physiology* 195 (2009): 989–1009; M. P. Nusbaum and D. M. Blitz, "Neuropeptide modulation of microcircuits," *Current Opinion in Neurobiology* 22 (2012): 592–601.

4 A. I. Selverston, "Are central pattern generators understandable?," *Behavioral and Brain Sciences* 3 (1980): 535–571.

5 J. P. Miller and A. Selverston, "Rapid killing of single neurons by irradiation of intracellularly injected dye," *Science* 206 (1979): 702–704; A. I. Selverston and J. P. Miller, "Mechanisms underlying pattern generation in lobster stomatogastric ganglion as determined by selective inactivation of identified neurons: I. Pyloric system," *Journal of Neurophysiology* 44 (1980): 1102–1121.

6 Selverston and Miller, "Mechanisms underlying pattern generation: I."

7 J. P. Miller and A. I. Selverston, "Mechanisms underlying pattern generation in lobster stomatogastric ganglion as determined by selective inactivation of identified neurons: IV. Network properties of pyloric system," *Journal of Neurophysiology* 48 (1982): 1416–1432.

8 A. A. Sharp et al., "The dynamic clamp: Artificial conductances in biological neurons," *Trends in Neurosciences* 16 (1993): 389–394; A. A. Sharp et al., "Dynamic clamp: Computer-generated conductances in real neurons," *Journal of Neurophysiology* 69 (1993): 992–995; A. A. Prinz et al., "The dynamic clamp comes of age," *Trends in Neurosciences* 27 (2004): 218–224.

9 P. S. Dickinson, "Neuromodulation of central pattern generators in invertebrates and vertebrates," *Current Opinion in Neurobiology* 16 (2006): 604–614; R. M. Harris-Warrick and B. R. Johnson, "Checks and balances in neuromodulation," *Frontiers in Behavioral Neuroscience* 4 (2010): article 47, doi: 10.3389/fnbeh.2010.00047; R. M. Harris-Warrick, "Neuromodulation and flexibility in Central Pattern Generator networks," *Current Opinion in Neurobiology* 21 (2011): 685–692; E. Marder, "Variability, compensation, and modulation in neurons and circuits," *Proceedings of the National Academy of Science USA* 108 Suppl 3 (2011): 15542–15548; E. Marder, "Neuromodulation of neuronal circuits: Back to the future," *Neuron* 76 (2012): 1–11.

10 Harris-Warrick, "Neuromodulation and flexibility."

11 D. J. Schulz et al., "Variable channel expression in identified single and electrically coupled neurons in different animals," *Nature Neuroscience* 9 (2006): 356–362; Marder, "Variability, compensation, and modulation"; E. Marder et al., "Robust circuit rhythms in small cir-

cuits arise from variable circuit components and mechanisms," *Current Opinion in Neurobiology* 31 (2015): 156–163.

12 A. A. Prinz et al., "Similar network activity from disparate circuit parameters," *Nature Neuroscience* 7 (2004): 1345–1352.

13 G. Turrigiano et al., "Activity-dependent changes in the intrinsic properties of cultured neurons," *Science* 264 (1994): 974–977.

14 G. Turrigiano, "Too many cooks? Intrinsic and synaptic homeostatic mechanisms in cortical circuit refinement," *Annual Review of Neuroscience* 34 (2011): 89–103.

15 J. N. MacLean et al., "Activity-independent homeostasis in rhythmically active neurons," *Neuron* 37 (2003): 109–120.

16 Ibid.; J. N. MacLean et al., "Activity-independent coregulation of IA and Ih in rhythmically active neurons," *Journal of Neurophysiology* 94 (2005): 3601–3617.

17 MacLean et al., "Activity-independent homeostasis"; MacLean et al., "Activity-independent coregulation."

18 L. S. Tang et al., "Precise temperature compensation of phase in a rhythmic motor pattern," *PLoS Biology* 8 (2010).

19 K. L. Briggman and W. B. Kristan, "Multifunctional pattern-generating circuits," *Annual Review of Neuroscience* 31 (2008): 271–294.

20 On sea slugs, see I. R. Popescu and W. N. Frost, "Highly dissimilar behaviors mediated by a multifunctional network in the marine mollusk *Tritonia diomedea*," *Journal of Neuroscience* 22 (2002): 1985–1993; on leeches, see K. L. Briggman and W. B. Kristan Jr., "Imaging dedicated and multifunctional neural circuits generating distinct behaviors," *Journal of Neuroscience* 26 (2006): 10925–10933; on sea hares, see J. Jing and K. R. Weiss, "Neural mechanisms of motor program switching in *Aplysia*," *Journal of Neuroscience* 21 (2001): 7349–7362; on tadpoles, see S. R. Soffe, "Two distinct rhythmic motor patterns are driven by common premotor and motor neurons in a simple vertebrate spinal cord," *Journal of Neuroscience* 13 (1993): 4456–4469; W. C. Li et al., "Reconfiguration of a vertebrate motor network: Specific neuron recruitment and context-dependent synaptic plasticity," *Journal of Neuroscience* 27 (2007): 12267–12276; on larval zebrafish, see D. A. Ritter et al., "In vivo imaging of zebrafish reveals differences in the spinal networks for escape and swimming movements," *Journal of Neuroscience* 21 (2001): 8956–8965; J. C. Liao and J. R. Fetcho, "Shared versus specialized glycinergic spinal interneurons in axial motor circuits of larval zebrafish," *Journal of Neuroscience* 28 (2008): 12982–12992; on turtles, see A. Berkowitz, "Both shared and specialized spinal circuitry for scratching and swimming

in turtles," *Journal of Comparative Physiology A—Sensory, Neural, and Behavioral Physiology* 188 (2002): 225–234; A. Berkowitz, "Multifunctional and specialized spinal interneurons for turtle limb movements," *Annals of the New York Academy of Sciences* 1198 (2010): 119–132; on cats, see M. B. Berkinblit et al., "Generation of scratching: II. Nonregular regimes of generation," *Journal of Neurophysiology* 41 (1978): 1058–1069; S. S. Geertsen et al., "Reciprocal Ia inhibition contributes to motoneuronal hyperpolarisation during the inactive phase of locomotion and scratching in the cat," *Journal of Physiology* 589 (2011): 119–134; A. Frigon, "Central pattern generators of the mammalian spinal cord," *Neuroscientist* 18 (2012): 56–69; on breathing in mammals, see S. P. Lieske et al., "Reconfiguration of the neural network controlling multiple breathing patterns: Eupnea, sighs and gasps," *Nature Neuroscience* 3 (2000): 600–607; A. Berkowitz et al., "Roles for multifunctional and specialized spinal interneurons during motor pattern generation in tadpoles, zebrafish larvae, and turtles," *Frontiers in Behavioral Neuroscience* 4 (2010): article 36, doi: 10.3389/fnbeh.2010.00036.

21 On leeches, see B. K. Shaw and W. B. Kristan Jr., "The neuronal basis of the behavioral choice between swimming and shortening in the leech: Control is not selectively exercised at higher circuit levels," *Journal of Neuroscience* 17 (1997): 786–795; on larval zebrafish, see Ritter et al., "In vivo imaging of zebrafish"; on turtles, see Berkowitz, "Multifunctional and specialized spinal interneurons."

22 Berkowitz et al., "Roles for multifunctional and specialized spinal interneurons."

23 On locusts, see J. M. Ramirez and K. G. Pearson, "Generation of motor patterns for walking and flight in motoneurons supplying bifunctional muscles in the locust," *Journal of Neurobiology* 19 (1988): 257–282; on crickets, see R. M. Hennig, "Neuronal control of the forewings in two different behaviours: Stridulation and flight in the cricket, *Teleogryllus commodus*," *Journal of Comparative Physiology A—Sensory, Neural, and Behavioral Physiology* 167 (1990): 617–627.

24 Berkowitz et al., "Roles for multifunctional and specialized spinal interneurons."

25 Briggman and Kristan, "Imaging dedicated and multifunctional neural circuits."

26 Jing and Weiss, "Neural mechanisms of motor program switching in *Aplysia*."

27 Liao and Fetcho, "Shared versus specialized glycinergic spinal interneurons"; Berkowitz et al., "Roles for multifunctional and specialized spinal interneurons."

28 C. Satou et al., "Functional role of a specialized class of spinal commissural inhibitory neurons during fast escapes in zebrafish," *Journal of Neuroscience* 29 (2009): 6780-6793.

29 Berkowitz, "Multifunctional and specialized spinal interneurons."

30 Berkowitz, "Both shared and specialized spinal circuitry"; A. Berkowitz, "Physiology and morphology of shared and specialized spinal interneurons for locomotion and scratching," *Journal of Neurophysiology* 99 (2008): 2887-2901.

31 A. Berkowitz, "Spinal interneurons that are selectively activated during fictive flexion reflex," *Journal of Neuroscience* 27 (2007): 4634-4641.

32 J. G. Puhl and K. A. Mesce, "Dopamine activates the motor pattern for crawling in the medicinal leech," *Journal of Neuroscience* 28 (2008): 4192-4200.

33 H. J. Rhodes et al., "*Xenopus* vocalizations are controlled by a sexually differentiated hindbrain central pattern generator," *Journal of Neuroscience* 27 (2007): 1485-1497.

34 J. R. Cazalets et al., "Two types of motor rhythm induced by NMDA and amines in an in vitro spinal cord preparation of neonatal rat," *Neuroscience Letters* 111 (1990): 116-121; J. R. Cazalets et al., "Activation of the central pattern generators for locomotion by serotonin and excitatory amino acids in neonatal rat," *Journal of Physiology* 455 (1992): 187-204; P. Whelan et al., "Properties of rhythmic activity generated by the isolated spinal cord of the neonatal mouse," *Journal of Neurophysiology* 84 (2000): 2821-2833.

35 P. S. Katz et al., "Dynamic neuromodulation of synaptic strength intrinsic to a central pattern generator circuit," *Nature* 367 (1994): 729-731.

36 T. Esch et al., "Evidence for sequential decision making in the medicinal leech," *Journal of Neuroscience* 22 (2002): 11045-11054.

37 K. L. Briggman et al., "Optical imaging of neuronal populations during decision-making," *Science* 307 (2005): 896-901.

38 A. Roberts et al., "How neurons generate behavior in a hatchling amphibian tadpole: An outline," *Frontiers in Behavioral Neuroscience* 4 (2010): article 16, doi: 10.3389/fnbeh.2010.00016.

39 Li et al., "Reconfiguration of a vertebrate motor network."

40 W. C. Li et al., "Specific brainstem neurons switch each other into pacemaker mode to drive movement by activating NMDA receptors," *Journal of Neuroscience* 30 (2010): 16609-16620.

41 P. R. Moult et al., "Fast silencing reveals a lost role for reciprocal inhibition in locomotion," *Neuron* 77 (2013): 129-140.

7. GOVERNMENT SURVEILLANCE

1 R. S. Payne, "Acoustic location of prey by barn owls *(Tyto alba),*" *Journal of Experimental Biology* 54 (1971): 535–573.

2 E. I. Knudsen et al., "Receptive fields of auditory neurons in the owl," *Science* 198 (1977): 1278–1280; E. I. Knudsen and M. Konishi, "A neural map of auditory space in the owl," *Science* 200 (1978): 795–797; M. Konishi, "Coding of auditory space," *Annual Review of Neuroscience* 26 (2003): 31–55; M. Konishi, "Behavioral guides for sensory neurophysiology," *Journal of Comparative Physiology A—Sensory, Neural, and Behavioral Physiology* 192 (2006): 671–676.

3 C. E. Carr and M. Konishi, "Axonal delay lines for time measurement in the owl's brainstem," *Proceedings of the National Academy of Science USA* 85 (1988): 8311–8315; C. E. Carr and M. Konishi, "A circuit for detection of interaural time differences in the brain stem of the barn owl," *Journal of Neuroscience* 10 (1990): 3227–3246.

4 L. A. Jeffress, "A place theory of sound localization," *Journal of Comparative and Physiological Psychology* 41 (1948): 35–39.

5 W. E. Sullivan and M. Konishi, "Segregation of stimulus phase and intensity coding in the cochlear nucleus of the barn owl," *Journal of Neuroscience* 4 (1984): 1787–1799; T. Takahashi et al., "Time and intensity cues are processed independently in the auditory system of the owl," *Journal of Neuroscience* 4 (1984): 1781–1786.

6 E. I. Knudsen and M. Konishi, "Mechanisms of sound localization in the barn owl *(Tyto alba),*" *Journal of Comparative Physiology* 133 (1979): 13–21.

7 G. A. Manley et al., "A neural map of interaural intensity differences in the brain stem of the barn owl," *Journal of Neuroscience* 8 (1988): 2665–2676.

8 A. Moiseff and M. Konishi, "Neuronal and behavioral sensitivity to binaural time differences in the owl," *Journal of Neuroscience* 1 (1981): 40–48.

9 K. Saberi et al., "How do owls localize interaurally phase-ambiguous signals?," *Proceedings of the National Academy of Science USA* 95 (1998): 6465–6468.

10 H. Wagner et al., "Representation of interaural time difference in the central nucleus of the barn owl's inferior colliculus," *Journal of Neuroscience* 7 (1987): 3105–3116.

11 E. I. Knudsen, "Auditory and visual maps of space in the optic tectum of the owl," *Journal of Neuroscience* 2 (1982): 1177–1194.

12 E. I. Knudsen and P. F. Knudsen, "Sensitive and critical periods for visual calibration of sound localization by barn owls," *Journal of Neuroscience* 10 (1990): 222–232; E. I. Knudsen and M. S. Brainard, "Vi-

sual instruction of the neural map of auditory space in the developing optic tectum," *Science* 253 (1991): 85–87.

13 Knudsen and Knudsen, "Sensitive and critical periods for visual calibration."

14 B. A. Linkenhoker and E. I. Knudsen, "Incremental training increases the plasticity of the auditory space map in adult barn owls," *Nature* 419 (2002): 293–296.

15 J. F. Bergan et al., "Hunting increases adaptive auditory map plasticity in adult barn owls," *Journal of Neuroscience* 25 (2005): 9816–9820.

16 K. C. Catania, "The sense of touch in the star-nosed mole: From mechanoreceptors to the brain," *Philosophical Transactions of the Royal Society of London, Series B: Biological Sciences* 366 (2011): 3016–3025.

17 K. C. Catania and J. H. Kaas, "Organization of the somatosensory cortex of the star-nosed mole," *Journal of Comparative Neurology* 351 (1995): 549–567.

18 K. C. Catania and J. H. Kaas, "Somatosensory fovea in the star-nosed mole: Behavioral use of the star in relation to innervation patterns and cortical representation," *Journal of Comparative Neurology* 387 (1997): 215–233.

19 K. C. Catania and F. E. Remple, "Tactile foveation in the star-nosed mole," *Brain, Behavior, and Evolution* 63 (2004): 1–12.

20 D. R. Griffin, "Return to the magic well: Echolocation behavior of bats and responses of insect prey," *BioScience* 51 (2001): 555–556.

21 H. Hartridge, "The avoidance of objects by bats in their flight," *Journal of Physiology* 54 (1920): 54–57.

22 G. W. Pierce and D. R. Griffin, "Experimental determination of supersonic notes emitted by bats," *Journal of Mammalogy* 19 (1938): 454–455.

23 D. R. Griffin and R. Galambos, "The sensory basis of obstacle avoidance by flying bats," *Journal of Experimental Zoology* 86 (1941): 481–506; D. R. Griffin, "Bat sounds under natural conditions, with evidence for the echolocation of insect prey," *Journal of Experimental Zoology* 123 (1953): 435–466.

24 N. Suga, "Principles of auditory information-processing derived from neuroethology," *Journal of Experimental Biology* 146 (1989): 277–286; N. Ulanovsky and C. F. Moss, "What the bat's voice tells the bat's brain," *Proceedings of the National Academy of Science USA* 105 (2008): 8491–8498.

25 Suga, "Principles of auditory information-processing"; Ulanovsky and Moss, "What the bat's voice tells the bat's brain."

26 N. Suga and W. E. O'Neill, "Neural axis representing target range in the auditory cortex of the mustache bat," *Science* 206 (1979): 351–353.

27 J. T. Sanchez et al., "Glycinergic 'inhibition' mediates selective excitatory responses to combinations of sounds," *Journal of Neuroscience* 28 (2008): 80–90; J. J. Wenstrup et al., "Mechanisms of spectral and temporal integration in the mustached bat inferior colliculus," *Frontiers in Neural Circuits* 6 (2012): article 75, doi: 10.3389/fncir. 2012.00075.

28 T. G. Brown, "The intrinsic factors in the act of progression in the mammal," *Proceedings of the Royal Society of London* 84 (1911): 308–319.

29 Suga and O'Neill, "Neural axis representing target range."

30 H. Riquimaroux et al., "Cortical computational maps control auditory perception," *Science* 251 (1991): 565–568.

31 J. A. Simmons et al., "Discrimination of jittered sonar echoes by the echolocating bat, *Eptesicus fuscus:* The shape of target images in echolocation," *Journal of Comparative Physiology A—Sensory, Neural, and Behavioral Physiology* 167 (1990): 589–616.

32 N. Suga and P. H. Jen, "Disproportionate tonotopic representation for processing CF-FM sonar signals in the mustache bat auditory cortex," *Science* 194 (1976): 542–544; G. Schuller and G. Pollak, "Disproportionate frequency representation in the inferior colliculus of Doppler-compensating greater horseshoe bats: Evidence for an acoustic fovea," *Journal of Comparative Physiology* 132 (1979): 47–54; H. U. Schnitzler and A. Denzinger, "Auditory fovea and Doppler shift compensation: Adaptations for flutter detection in echolocating bats using CF-FM signals," *Journal of Comparative Physiology A—Sensory, Neural, and Behavioral Physiology* 197 (2011): 541–559.

33 N. Suga et al., "Cortical neurons sensitive to combinations of information-bearing elements of biosonar signals in the mustache bat," *Science* 200 (1978): 778–781.

34 L. H. Goldman and O. W. Henson, "Prey recognition and selection by the constant frequency bat, *Pteronotus p. parnelli,*" *Behavioral Ecology and Sociobiology* 2 (1977): 411–419.

35 K. D. Roeder, *Nerve Cells and Insect Behavior* (Cambridge, MA: Harvard University Press, 1967).

36 Ibid.; A. J. Corcoran et al., "Tiger moth jams bat sonar," *Science* 325 (2009): 325–327.

37 C. Chiu et al., "Flying in silence: Echolocating bats cease vocalizing to avoid sonar jamming," *Proceedings of the National Academy of Science USA* 105 (2008): 13116–13121.

38 D. R. Griffin, *Listening in the Dark: The Acoustic Orientation of Bats and Men* (New Haven, CT: Yale University Press, 1958).

39 Chiu et al., "Flying in silence."

40 A. J. Corcoran and W. E. Conner, "Bats jamming bats: Food competition through sonar interference," *Science* 346 (2014): 745-747.

41 H. W. Lissmann, "Continuous electrical signals from the tail of a fish, *Gymnarchus niloticus* Cuv.," *Nature* 167 (1951): 201-202.

42 A. Watanabe and K. Takeda, "The change of discharge frequency by AC stimulus in a weak electric fish," *Journal of Experimental Biology* 40 (1963): 57-66; G. J. Rose, "Insights into neural mechanisms and evolution of behaviour from electric fish," *Nature Reviews Neuroscience* 5 (2004): 943-951; Konishi, "Behavioral guides for sensory neurophysiology"; G. K. Zupanc and T. H. Bullock, "Walter Heiligenberg: The jamming avoidance response and beyond," *Journal of Comparative Physiology A—Sensory, Neural, and Behavioral Physiology* 192 (2006): 561-572.

43 G. Rose and W. Heiligenberg, "Temporal hyperacuity in the electric sense of fish," *Nature* 318 (1985): 178-180.

8. GOVERNMENT SELF-MONITORING

1 H. L. Helmholtz, *Handbuch der Physiologischen Optik* [Treatise on physiological optics], vol. 3: *The Perceptions of Vision*, trans. J. P. C. Southall (Hamburg: Verlag von Leopold Voss, 1910).

2 R. W. Sperry, "Neural basis of the spontaneous optokinetic response produced by visual inversion," *Journal of Comparative and Physiological Psychology* 43 (1950): 482-489; E. von Holst and H. Mittlestaedt, "Das reafferenzprincip" [The reafference principle], *Naturwissenschaften* 37 (1950): 464-476, in *The Collected Papers of Erich von Holst*, vol. 1: *The Behavioural Physiology of Animals and Man*, trans. Robert Martin (Coral Gables: University of Miami Press, 1973).

3 M. A. Sommer and R. H. Wurtz, "A pathway in primate brain for internal monitoring of movements," *Science* 296 (2002): 1480-1482.

4 F. Huber, "Nerve cells and insect behavior-studies on crickets," *American Zoologist* 30 (1990): 609-627; B. Hedwig, "Pulses, patterns and paths: Neurobiology of acoustic behaviour in crickets," *Journal of Comparative Physiology A—Sensory, Neural, and Behavioral Physiology* 192 (2006): 677-689.

5 J. F. Poulet and B. Hedwig, "A corollary discharge maintains auditory sensitivity during sound production," *Nature* 418 (2002): 872-876; J. F. Poulet and B. Hedwig, "The cellular basis of a corollary discharge," *Science* 311 (2006): 518-522.

6 On other animal behaviors generally, see T. B. Crapse and M. A. Sommer, "Corollary discharge across the animal kingdom," *Nature Reviews Neuroscience* 9 (2008): 587–600; on mollusk feeding, see M. P. Kovac and W. J. Davis, "Behavioral choice: Neural mechanisms in *Pleurobranchaea*," *Science* 198 (1977): 632–634; M. P. Kovac et al., "Food avoidance learning is accompanied by synaptic attenuation in identified interneurons controlling feeding behavior in *Pleurobranchaea*," *Journal of Neurophysiology* 56 (1986): 891–905; on crayfish escape, see M. D. Kirk, "Presynaptic inhibition in the crayfish CNS: Pathways and synaptic mechanisms," *Journal of Neurophysiology* 54 (1985): 1305–1325; on crayfish walking, see K. T. Sillar and P. Skorupski, "Central input to primary afferent neurons in crayfish, *Pacifastacus leniusculus*, is correlated with rhythmic motor output of thoracic ganglia," *Journal of Neurophysiology* 55 (1986): 678–688; D. Cattaert et al., "Direct evidence for presynaptic inhibitory mechanisms in crayfish sensory afferents," *Journal of Neurophysiology* 67 (1992): 610–624; F. Clarac et al., "Central control components of a 'simple' stretch reflex," *Trends in Neurosciences* 23 (2000): 199–208; on cockroach walking, see F. Delcomyn, "Corollary discharge to cockroach giant interneurones," *Nature* 269 (1977): 160–162; on locust head movements, see M. Zaretsky and C. H. Rowell, "Saccadic suppression by corollary discharge in the locust," *Nature* 280 (1979): 583–585; on electric fish electric organ discharges, see W. Heiligenberg and J. Bastian, "The electric sense of weakly electric fish," *Annual Review of Physiology* 46 (1984): 561–583; C. C. Bell et al., "Storage of a sensory pattern by anti-Hebbian synaptic plasticity in an electric fish," *Proceedings of the National Academy of Science USA* 90 (1993): 4650–4654; on fish calling, see B. P. Chagnaud et al., "Vocalization frequency and duration are coded in separate hindbrain nuclei," *Nature Communications* 2 (2011): 346; on skate ventilation, see G. Hjelmstad et al., "Motor corollary discharge activity and sensory responses related to ventilation in the skate vestibulolateral cerebellum: Implications for electrosensory processing," *Journal of Experimental Biology* 199 (1996): 673–681; on tadpole swimming, see K. T. Sillar and A. Roberts, "A neuronal mechanism for sensory gating during locomotion in a vertebrate," *Nature* 331 (1988): 262–265; on zebra finch singing, see G. F. Striedter and E. T. Vu, "Bilateral feedback projections to the forebrain in the premotor network for singing in zebra finches," *Journal of Neurobiology* 34 (1998): 27–40; R. Mooney and J. F. Prather, "The HVC microcircuit: The synaptic basis for interactions between song motor and vocal plasticity pathways," *Journal of Neuroscience* 25 (2005): 1952–1964; on rat whisker move-

ments, see M. S. Fee et al., "Central versus peripheral determinants of patterned spike activity in rat vibrissa cortex during whisking," *Journal of Neurophysiology* 78 (1997): 1144–1149; on cat walking, see Y. I. Arshavsky et al., "Origin of modulation in neurones of the ventral spinocerebellar tract during locomotion," *Brain Research* 43 (1972): 276–279; P. Rudomin and R. F. Schmidt, "Presynaptic inhibition in the vertebrate spinal cord revisited," *Experimental Brain Research* 129 (1999): 1–37; D. A. McCrea, "Spinal circuitry of sensorimotor control of locomotion," *Journal of Physiology* 533 (2001): 41–50; on cat scratching, see Y. I. Arshavsky et al., "Messages conveyed by spinocerebellar pathways during scratching in the cat: II. Activity of neurons of the ventral spinocerebellar tract," *Brain Research* 151 (1978): 493–506; on human walking, see J. F. Yang and R. B. Stein, "Phase-dependent reflex reversal in human leg muscles during walking," *Journal of Neurophysiology* 63 (1990): 1109–1117; on human hand movements, see S. C. Gandevia et al., "Motor commands contribute to human position sense," *Journal of Physiology* 571 (2006): 703–710.

9. BECOMING A POLITICAL ANIMAL

1 M. Konishi, "Birdsong: From behavior to neuron," *Annual Review of Neuroscience* 8 (1985): 125–170; J. J. Bolhuis and M. Gahr, "Neural mechanisms of birdsong memory," *Nature Reviews Neuroscience* 7 (2006): 347–357; R. Mooney, "Neural mechanisms for learned birdsong," *Learning & Memory* 16 (2009): 655–669; M. S. Brainard and A. J. Doupe, "Translating birdsong: Songbirds as a model for basic and applied medical research," *Annual Review of Neuroscience* 36 (2013): 489–517.

2 Konishi, "Birdsong."

3 Ibid.

4 P. K. Kuhl, "Brain mechanisms in early language acquisition," *Neuron* 67 (2010): 713–727.

5 Konishi, "Birdsong."

6 Ibid.

7 A. C. Yu and D. Margoliash, "Temporal hierarchical control of singing in birds," *Science* 273 (1996): 1871–1875.

8 E. T. Vu et al., "Identification of a forebrain motor programming network for the learned song of zebra finches," *Journal of Neuroscience* 14 (1994): 6924–6934.

9 A. S. Andalman et al., "Control of vocal and respiratory patterns in birdsong: Dissection of forebrain and brainstem mechanisms using temperature," *PLoS One* 6 (2011): e25461.

10 F. Nottebohm et al., "Central control of song in the canary, *Serinus canarius*," *Journal of Comparative Neurology* 165 (1976): 457–486.

11 Yu and Margoliash, "Temporal hierarchical control of singing in birds"; R. H. Hahnloser et al., "An ultra-sparse code underlies the generation of neural sequences in a songbird," *Nature* 419 (2002): 65–70.

12 Vu et al., "Identification of a forebrain motor programming network."

13 Hahnloser et al., "Generation of neural sequences in a songbird."

14 J. S. McCasland and M. Konishi, "Interaction between auditory and motor activities in an avian song control nucleus," *Proceedings of the National Academy of Science USA* 78 (1981): 7815–7819.

15 D. Margoliash, "Acoustic parameters underlying the responses of song-specific neurons in the white-crowned sparrow," *Journal of Neuroscience* 3 (1983): 1039–1057.

16 E. S. Fortune et al., "Neural mechanisms for the coordination of duet singing in wrens," *Science* 334 (2011): 666–670.

17 McCasland and Konishi, "Interaction between auditory and motor activities"; J. F. Prather et al., "Precise auditory-vocal mirroring in neurons for learned vocal communication," *Nature* 451 (2008): 305–310.

18 Prather et al., "Precise auditory-vocal mirroring."

19 Ibid.; C. Keysers and V. Gazzola, "Expanding the mirror: Vicarious activity for actions, emotions, and sensations," *Current Opinion in Neurobiology* 19 (2009): 666–671; G. Rizzolatti and C. Sinigaglia, "The functional role of the parieto-frontal mirror circuit: Interpretations and misinterpretations," *Nature Reviews Neuroscience* 11 (2010): 264–274; A. Casile, "Mirror neurons (and beyond) in the macaque brain: An overview of 20 years of research," *Neuroscience Letters* 540 (2013): 3–14.

20 Keysers and Gazzola, "Expanding the mirror."

21 C. Scharff and F. Nottebohm, "A comparative study of the behavioral deficits following lesions of various parts of the zebra finch song system: Implications for vocal learning," *Journal of Neuroscience* 11 (1991): 2896–2913.

22 A. J. Doupe and M. Konishi, "Song-selective auditory circuits in the vocal control system of the zebra finch," *Proceedings of the National Academy of Science USA* 88 (1991): 11339–11343.

23 K. W. Nordeen and E. J. Nordeen, "Auditory feedback is necessary for the maintenance of stereotyped song in adult zebra finches," *Behavioral and Neural Biology* 57 (1992): 58–66.

24 A. Leonardo and M. Konishi, "Decrystallization of adult birdsong by perturbation of auditory feedback," *Nature* 399 (1999): 466–470.

25 M. S. Brainard and A. J. Doupe, "Interruption of a basal ganglia-forebrain circuit prevents plasticity of learned vocalizations," *Nature* 404 (2000): 762–766.

26 M. H. Kao et al., "Contributions of an avian basal ganglia-forebrain circuit to real-time modulation of song," *Nature* 433 (2005): 638–643; B. P. Olveczky et al., "Vocal experimentation in the juvenile songbird requires a basal ganglia circuit," *PLoS Biology* 3 (2005): e153.

27 M. H. Kao et al., "Neurons in a forebrain nucleus required for vocal plasticity rapidly switch between precise firing and variable bursting depending on social context," *Journal of Neuroscience* 28 (2008): 13232–13247.

28 E. C. Tumer and M. S. Brainard, "Performance variability enables adaptive plasticity of 'crystallized' adult birdsong," *Nature* 450 (2007): 1240–1244; S. J. Sober and M. S. Brainard, "Adult birdsong is actively maintained by error correction," *Nature Neuroscience* 12 (2009): 927–931.

29 A. S. Andalman and M. S. Fee, "A basal ganglia-forebrain circuit in the songbird biases motor output to avoid vocal errors," *Proceedings of the National Academy of Science USA* 106 (2009): 12518–12523.

30 J. D. Charlesworth et al., "Covert skill learning in a cortical-basal ganglia circuit," *Nature* 486 (2012): 251–255.

31 F. Ali et al., "The basal ganglia is necessary for learning spectral, but not temporal, features of birdsong," *Neuron* 80 (2013): 494–506.

32 J. E. Markowitz et al., "Mesoscopic patterns of neural activity support songbird cortical sequences," *PLoS Biology* 13 (2015): e1002158, doi:10.1371/journal.pbio.1002158.

33 G. B. Keller and R. H. Hahnloser, "Neural processing of auditory feedback during vocal practice in a songbird," *Nature* 457 (2009): 187–190.

10. GOVERNING BEHAVIOR

1 E. Marder and D. Bucher, "Understanding circuit dynamics using the stomatogastric nervous system of lobsters and crabs," *Annual Review of Physiology* 69 (2007): 291–316; W. Stein, "Modulation of stomato-gastric rhythms," *Journal of Comparative Physiology A—Sensory, Neural, and Behavioral Physiology* 195(2009): 989–1009.

2 J. L. Feldman et al., "Understanding the rhythm of breathing: So near, yet so far," *Annual Review of Physiology* 75 (2013): 423–452; J. L. Feldman and K. Kam, "Facing the challenge of mammalian neural microcircuits: Taking a few breaths may help," *Journal of Physiology* 593 (2014): 3–23.

3 E. Marder, "Variability, compensation, and modulation in neurons and circuits," *Proceedings of the National Academy of Science USA* 108 Suppl 3 (2011): 15542–15548.

4 G. Turrigiano, "Homeostatic synaptic plasticity: Local and global mechanisms for stabilizing neuronal function," *Cold Spring Harbor Perspectives in Biology* 4 (2012): a005736.

5 "How a trading algorithm went awry," *Wall Street Journal*, October 2, 2010.

6 Marder, "Variability, compensation, and modulation."

Credits

Figure 1.1: Illustration by the author, from the author's data.

Figure 1.2: Composite adapted by permission of Oxford University Press, USA, from Santiago Ramon y Cajal, *Histologie du Système Nerveux de l'homme & des Vertébrés* (1909), Maloine: Paris, as taken from *Histology of the Nervous System*, vol. 1, trans. N. Swanson and L. W. Swanson. 1995. Figs. 8, 9, 11, and 401; pp. 53–55, 741. Copyright © 1995 by Oxford University Press, Inc.

Figure 2.1: Illustration by the author.

Figure 2.2: *(A)* Reprinted by permission from Macmillan Publishers Ltd.: S. E. Petersen, P. T. Fox, M. I. Posner, M. Minton, and M. E. Raichle. 1998. Positron emission tomographic studies of the cortical anatomy of single-word processing. *Nature* 331: 585–589. Fig. 2(g). Copyright © 1988 *Nature* Publishing Group. Image courtesy of Steven Petersen, Washington University. *(B)* Reprinted with permission from AAAS from A. R. Hariri, V. S. Mattay, A. Tessitore, B. Kolachana, F. Fera, D. Goldman, M. F. Egan, and D. R. Weinberger. 2002. Serotonin transporter genetic variation and the response of the human amygdala. *Science* 297: 400–403. Fig. 2, Second Cohort. Copyright © 2002 by the American Association for the Advancement of Science.

Figure 3.1: Photographs taken and provided by Jens Herberholz and Abigail Schadegg, University of Maryland, College Park.

Figure 3.2: Reprinted by permission of Elsevier from J. R. Fetcho and D. L. McLean. 2009. Startle response. In *Encyclopedia of Neuroscience*, ed. L. R. Squire, 375–379. Fig. 2. Copyright © 2009 Elsevier Ltd.

Figure 3.3: Illustration by Coral McCallister.

Figure 4.1: Illustration by Coral McCallister.

Figure 4.2: Reproduced from J. J. Heys, P. K. Rajaraman, T. Gedeon, and J. P. Miller. 2012. A model of filiform hair distribution on the cricket cercus. *PLoS ONE* 7(10): e46588. doi:10.1371/journal.pone.0046588. Fig. 1. Copyright © 2012 Heys et al. Creative Commons Attribution 4.0 International Public License.

Figure 4.3: *(Above)* Images of neurons courtesy of John P. Miller, Montana State University. *(Below)* Reprinted by permission of the American Physiological Society (APS) from J. P. Miller, G. A. Jacobs, and F. E. Theunissen. 1991. Representation of sensory information in the cricket cercal sensory system. I. Response properties of the primary interneurons. *Journal of Neurophysiology* 66: 1680–1689. Fig. 5(A), p. 1685. Copyright © 1991 the American Physiological Society (APS).

Figure 4.4: Reprinted with permission from Elsevier from W. B. Kristan Jr., R. L. Calabrese, and W. O. Friesen. 2005. Neuronal control of leech behavior. *Progress in Neurobiology* 76: 279–327. Fig. 3(A). Copyright © 2005 Elsevier Ltd. Image courtesy of William B. Kristan Jr., University of California, San Diego.

Figure 4.5: Reprinted by permission from Macmillan Publishers Ltd.: J. E. Lewis and W. B. Kristan Jr. 1998. A neuronal network for computing population vectors in the leech. *Nature* 391: 76–79. Fig. 2. Copyright © 1998 Macmillan Publishers Ltd.

Figure 4.6: Reprinted from D. L. Sparks, R. Holland, and B. L Guthrie. 1976. Size and distribution of movement fields in the monkey superior colliculus. *Brain Research* 113 (1976): 21–34. Fig. 5. © Copyright Elsevier Scientific Publishing Company, Amsterdam, 1976. Reproduction prepared by Coral McCallister.

Figure 4.7: Reprinted with permission from AAAS from A. P. Georgopoulos, A. B. Schwartz, and R. E. Kettner. 1986. Neuronal population coding of movement direction. *Science* 233: 1416–1419. Fig. 3. Copyright © 1986 American Association for the Advancement of Science. Prepared for black and white reproduction by Coral McCallister.

Figure 4.8: Reprinted by permission from Macmillan Publishers Ltd.: C. T. Moritz, S. I. Perlmutter, and E. E. Fetz. 2008. Direct control of paralysed muscles by cortical neurons. *Nature* 456: 639–642. Fig. 1. Copyright © 2008 Macmillan Publishers Ltd. Image courtesy of Chet Moritz, University of Washington. Prepared for black and white reproduction by Coral McCallister.

Figure 5.1: Illustration by the author.

Figure 5.2: Reproduced from T. Graham Brown. 1911. The intrinsic factors in the act of progression in the mammal. *Proceedings of the Royal Society of London* 84: 308–319. Fig. 1 (detail). Copyright © 1911 by the Royal Society.

Figure 5.3: Composite reprinted from C. A. Del Negro, J. A. Hayes, R. W. Pace, B. R. Brush, R. Teruyama, and J. L. Feldman. 2010. Synaptically activated burst-generating conductances may underlie a group-pacemaker mechanism for respiratory rhythm generation in mammals. *Progress in Brain Research* 187: 111–136. Figs. 1(C) and 4(A). Copyright © 2010 Elsevier B.V., with permission from Elsevier.

Figure 6.1: Illustration by Coral McCallister.

Figure 6.2: Reprinted with permission from A. A. Sharp, F. K. Skinner, and E. Marder. 1996. Mechanisms of oscillation in dynamic clamp constructed two-cell half-center circuits. *Journal of Neurophysiology* 76: 867–883. Fig. 3(E). Copyright © 1996 by the American Physiological Society (APS). Image courtesy of Eve Marder, Brandeis University.

Figure 6.3: Illustration by the author, from the author's data.

Figure 6.4: Reprinted with permission from Elsevier from W. B. Kristan Jr., R. L. Calabrese, and W. O. Friesen. 2005. Neuronal control of leech behavior. *Progress in Neurobiology* 76: 279–327. Figs. 3(B) and (D). Copyright © 2005 Elsevier Ltd.

Figure 6.5: Reprinted from A. Roberts, W.-C. Li, and S. R. Soffe. 2010. How neurons generate behavior in a hatchling amphibian tadpole: An outline. *Frontiers in Behavioral Neuroscience* 4: 16. Figs. 1(D) and (F). Copyright © 2010 Roberts, Li and Soffe. Creative Commons Attribution 4.0 International Public License.

Figure 6.6: Republished with permission of the Society for Neuroscience from W.-C. Li, B. Sautois, A. Roberts, and S. R. Soffe. 2007. Reconfiguration of a vertebrate motor network: Specific neuron recruitment and context-dependent synaptic plasticity. *Journal of Neuroscience* 27: 12267–12276. Fig. 8. Copyright © 2007 Society for Neuroscience. Image courtesy of Alan Roberts. Prepared for black and white reproduction by Coral McCallister.

Figure 7.1: Photograph courtesy of Tony Hisgett, Birmingham, UK (Female Barn Owl 1, 6796240760), Flickr. Creative Commons Attribution 2.0 Generic License.

Figure 7.2: Photograph courtesy of Masakazu Konishi, California Institute of Technology, conveyed by Eric Knudsen, Stanford University.

Figure 7.3: Illustration by the author.

Figure 7.4: Reprinted with permission from Elsevier from K. C. Catania. 2005. Star-nosed moles. *Current Biology* 15: R863– R864. Copyright © 2005 Elsevier Ltd. Images courtesy of Kenneth Catania, Vanderbilt University.

Figure 7.5: Photograph courtesy of Roberto Leonan Morim Novaes, Federal University of Rio de Janeiro.

Figure 7.6: Illustration by the author.

Figure 7.7: Adapted with permission from N. Matsuta, S. Hiryu, E. Fujioka, Y. Yamada, H. Riquimaroux, and Y. Watanabe. 2013. Adaptive beam-width control of echolocation sounds by CF–FM bats, *Rhinolophus ferrumequinum nippon,* during prey-capture flight. *Journal of Experimental Biology* 216: 1210–1218. Fig. 2, above. Copyright © 2013. Published by the Company of Biologists Ltd. Black and white image courtesy of Shizuko Hiryu, Doshisha University. Animal drawings added by Coral McCallister.

Figure 7.8: Reprinted with permission from H. W. Lissmann and K. E. Machin. 1958. The mechanism of object location in *Gymnarchus niloticus* and similar fish. *Journal of Experimental Biology* 35: 451–486. Fig. 2. Copyright © 1958 by the Company of Biologists Ltd.

Figure 7.9: Courtesy of Carl Hopkins, Cornell University, and Masashi Kawasaki, University of Virginia.

Figure 7.10: Illustration by Coral McCallister.

Figure 8.1: Adapted with kind permission from Springer Science and Business Media from E. von Holst and H. Mittelstaedt. 1950. Das Reafferenzprinzip. Wedlselwirkungen zwischen Zentrainervensystem und Peripherie. *Naturwissenschaften* 37: 464–476. Fig. 5a–d, p. 469. Copyright © 1950 Springer-Verlag.

Figure 8.2: Composite reprinted with permission from AAAS from J. F. A. Poulet and B. Hedwig. 2006. The cellular basis of a corollary discharge. *Science* 311: 518–522. Figs. 2(A), (C), (G) and 3(D), (E). Copyright © 2006 by the American Association for the Advancement of Science. Images courtesy of James Poulet, Max Delbrück Center for Molecular Medicine, Berlin.

Figure 9.1: Photograph courtesy of Keith Gerstung from McHenry, IL *(Taeniopygia guttata),* Flickr via Wikimedia Commons. Creative Commons Attribution 2.0 Generic License.

Figure 9.2: Illustration by Coral McCallister.

Figure 9.3: Reprinted by permission from Macmillan Publishers Ltd. from R. H. R. Hahnloser, A. A. Kozhevnikov, and M. S. Fee. 2002. An ultra-sparse code underlies the generation of neural sequences in a songbird. *Nature* 419: 65–70. Fig. 2(B). Copyright © 2002 *Nature* Publishing Group. Image courtesy of Michale Fee, Massachusetts Institute of Technology.

Acknowledgments

I am especially grateful to colleagues, friends, and family members who took the time to read and provide feedback on part or all of this book in its earlier versions: Avery Berkowitz, Nisan Chavkin, Marshall Cheney, Betsy Dobbins, Julie Fiez, Rani Levinson, John Miller, Peter Narins, Benjamin White, and anonymous peer reviewers.

I am indebted to neuroethology and neurobiology colleagues who generously and expeditiously provided their images for some of the book's illustrations, often on relatively short notice: Catherine Carr, Kenneth Catania, Michale Fee, Joe Fetcho, Berthold Hedwig, Jens Herberholz, Shizuko Hiryu, Carl Hopkins, Masashi Kawasaki, Eric Knudsen, Phyllis Knudsen, Mark Konishi, Bill Kristan, John Lewis, Eve Marder, John Miller, Chet Moritz, Roberto Leonan Morim Novaes, Jose Pena, James Poulet, Alan Roberts, Abigail Schadegg, and Terry Takahashi. Coral McCallister (coral-sartstudio@gmail.com) created several new illustrations and redrew several others, also on short notice.

I thank students in my Behavioral Neurobiology (previously called Neuroethology) and Neural Control of Movement courses over the years for stimulating discussions and providing the impetus to clarify my explanations of experiments described in this book. I thank Masashi Kawasaki and Gunther Zupanc for sharing a story about Walter Heiligenberg and Iya Prytkova for informing me about Gogol's short story.

I thank several people who gave me advice about publishing my first book, early in the process: Joan Bossert, Rob Dunn, Alison Kalett, Michael Mares, Doug Mock, Ingo Schlupp, Tom Seeley, Peter Tallack, and Don Wilson.

I thank the editors at Harvard University Press, especially Michael Fisher, who provided sustained support for this book, Thomas Embree LeBien (following Michael's retirement), Susan Wallace Boehmer, and Kate Brick, as well as the staff who helped guide me through the various stages of this project: Lauren Esdaile, Michael Higgins, Amanda Peery, Stephanie Vyce, and Anne Zarrella. I thank John Donohue and Ashley Moore at Westchester Publishing Services for careful copyediting.

Financial support for the final stages of this book was provided partly by the National Science Foundation (NSF), via Integrative Organismal Systems grant 1354522. Without the support of the NSF, there would be almost no neuroethology research in the United States today, not to mention science in a form that the general public can understand and appreciate.

Index

Acetylcholine, 81. *See also* Chemical synapse; Neurotransmitter; Receptor

Anterior forebrain pathway (AFP), 173–174, 176. *See also* Bird; Learning; Singing

Axon, 3–4, 9; on giant axons, 34; in crayfish, 42, 44; from leg, 84–86; in locusts, 91; in crustaceans, 97; in tadpoles, 115; in owls, 125–126; in monkeys, 159; in songbirds, 168, 170–173. *See also* Cell body; Dendrite

Azimuth, 124–129, 135, 138. *See also* Delay lines

Bat, 9, 68, 120–121, 134–142, 148. *See also* Echolocation

Bird, 10, 75; on songbirds, 165–178, 183. *See also* Owl; Zebra finch

Blood flow, 22–23, 25, 28–29, 119. *See also* Brain scan; Functional magnetic resonance imaging (fMRI); Positron emission tomography (PET)

Brain-machine interface (BMI), 69–71, 74

Brain scan, 12, 21–26, 28–29, 31. *See also* Functional magnetic resonance imaging (fMRI); Positron emission tomography (PET)

Brainstem, 92–93, 96, 110, 159. *See also* Medulla; Midbrain

Breathing (or breathe), 9, 75–76, 91–97, 100, 107, 117–119, 179–181. *See also* Brainstem; Medulla

Broad tuning (or broadly tuned), 58–60, 66, 181–182. *See also* Democracy; Election; Population coding; Vote

Calcium, 77–78, 80–81. *See also* Channel; Ion

Calling, 91, 110, 163. *See also* Singing

Cardinal cell, 47, 49. *See also* Grandmother cell; Hierarchy; Oligarchy; Pontifical; Sparse coding

Cat: on population coding, 68; on walking/scratching, 83, 85–86, 88, 94, 100, 163; on multifunctional neurons, 107

Cell body, 3–4. *See also* Axon; Dendrite

Central program (or central pattern generator), 9, 76, 82, 96, 118, 163, 179–180, 182; on cat walking/scratching, 88; on locust flying, 91; on mammalian breathing, 93; on crustacean digestion, 97–98; on birdsong, 166, 169

Cerebrum (or cerebral cortex), 29; on face recognition, 50; on star-nosed moles, 131, 133; on bats, 135, 138–140; on corollary discharge, 159. *See also* Frontal lobe; Motor cortex; Parietal lobe; Temporal lobe